Studies in Computational Intelligence

Volume 949

The series "Studies in Computational Intelligence" (SCI) publishes new developments and advances in the various areas of computational intelligence—quickly and with a high quality. The intent is to cover the theory, applications, and design methods of computational intelligence, as embedded in the fields of engineering, computer science, physics and life sciences, as well as the methodologies behind them. The series contains monographs, lecture notes and edited volumes in computational intelligence spanning the areas of neural networks, connectionist systems, genetic algorithms, evolutionary computation, artificial intelligence, cellular automata, self-organizing systems, soft computing, fuzzy systems, and hybrid intelligent systems. Of particular value to both the contributors and the readership are the short publication timeframe and the world-wide distribution, which enable both wide and rapid dissemination of research output.

Indexed by SCOPUS, DBLP, WTI Frankfurt eG, zbMATH, SCImago.

All books published in the series are submitted for consideration in Web of Science.

More information about this series at http://www.springer.com/series/7092

Martin Stettinger · Gerhard Leitner ·
Alexander Felfernig · Zbigniew W. Ras
Editors

Intelligent Systems in Industrial Applications

Springer

Editors
Martin Stettinger
Graz University of Technology
Graz, Austria

Gerhard Leitner
University of Klagenfurt
Klagenfurt, Austria

Alexander Felfernig
Graz University of Technology
Klagenfurt, Austria

Zbigniew W. Ras
University of North Carolina
Charlotte, NC, USA

ISSN 1860-949X ISSN 1860-9503 (electronic)
Studies in Computational Intelligence
ISBN 978-3-030-67147-1 ISBN 978-3-030-67148-8 (eBook)
https://doi.org/10.1007/978-3-030-67148-8

This Springer imprint is published by the registered company Springer Nature Switzerland AG
The registered company address is: Gewerbestrasse 11, 6330 Cham, Switzerland

Preface

This book represents a collection of papers selected for the Industrial Part of ISMIS 2020 (25th International Symposium on Methodologies for Intelligent Systems) which has virtually taken place in Graz, Austria, in 2020. ISMIS was organized by the Institute of Software Technology of the Technical University of Graz, Austria, in cooperation with the Institute of Informatics Systems of the University of Klagenfurt, Austria.

ISMIS is a conference series that started in 1986. Held twice every three years, it provides an international forum for exchanging scientific, research, and technological achievements in building intelligent systems. In particular, major areas selected for ISMIS 2020 include explainable AI (XAI), machine learning, deep learning, data mining, recommender systems, constraint-based systems, autonomous systems, applications (configuration, Internet of things, financial services, e-Health, …), intelligent user interfaces, user modeling, human computation, socially aware systems, autonomous systems, digital libraries intelligent agents, information retrieval, natural language processing, knowledge integration and visualization, knowledge representation, soft computing, and web and text mining.

Besides the scientifically oriented part of ISMIS, the organizers were happy to be able to establish a separate track of papers with a focus on application of research outcome, represented in this book. A broad range of application possibilities is characterizing the presented papers, which were carefully selected in a single-blind review process, where at least three anonymous reviewers evaluated the submissions. We as the organizers appreciate the broad range of applicability, because this is an indicator for the relevance of the work of our community.

The selection of papers starts with an example showing an application in the automotive sector. The paper of Elzbieta Kubera, Alicja Wieczorkowska, and Andrzej Kuranc deals with the possibilities of applying AI methods for the automation of economic driving. Apart from environmental benefits, road safety is also increased when drivers avoid speeding and sudden changes of speeds, but speed measurements usually do not include such information. The authors therefore focus their work on automatic detection of speed, whereas three classes are of relevance: accelerating, decelerating, and maintaining stable speed. Theoretical discussions of the thresholds for these classes are followed by experiments with an automatic search for these thresholds. The obtained results are emphasized in the paper.

A sector related to automotive, concretely transportation, and logistics is the topic of the paper presented by Marie Le Guilly, Claudia Capo, Jean-Marc Petit, Marian Scuturici, Rémi Revellin, Jocelyn Bonjour, and Gérald Cavalier. The authors apply machine learning methods to predict aging and durability of vehicles transporting refrigerated goods. They focus their work on the company CEMAFROID, a french delegated public service, delivering conformity attestations of refrigerated transport vehicles. The DATAFRIG database opens the opportunity to predict the aging, however with some limitations the authors emphasize in their paper. They propose to use the notion of functional dependencies to address these limitations. The approach has been evaluated with domain experts from CEMAFROID, with many positive feedbacks.

The next selection of papers addresses—in a broader sense—aspects of learning. Stefano Ferilli, Giovanni Luca Izzi, and Tiziano Franza focus their work on natural language processing and present an attempt to automatically derive tools to support natural language processing. Such tools are useful linguistic resources, but not available for many languages. Since manually building them would be complex, the authors emphasize ways to generate such tools automatically, for example from sample texts. In their paper, the authors focus on stopwords, i.e., terms which are not relevant to understand the topic and content of a document and investigate other techniques proposed in the literature. The basic language investigated is Italian, and the presented approach is generic and applicable to other languages, too.

Azim Roussanaly, Marharyta Aleksandrova, and Anne Boyer focus their work on students who failed the final examination in the secondary school in France (known as baccalauréat or baccalaureate). In this case, students can improve their scores by passing a remedial test. This test consists of two oral examinations in two subjects of the student's choice. Students announce their choice on the day of the remedial test. However, the secondary education system in France is quite complex. There exist several types of baccalaureate consisting of various streams. Depending upon the stream students belong to, they have different subjects allowed to be taken during the remedial test and different coefficients associated with each of them. The authors present BacAnalytics—a tool that was developed to assist the rectorate of secondary schools with the organization of remedial tests for the baccalaureate.

Anna Saranti, Simon Streit, Heimo Müller, Deepika Singh, and Andreas Holzinger investigate visual concept learning methodologies which have become the state-of-the-art research that challenges the reasoning capabilities of deep learning methods. In their paper, the authors discuss the evolution of those methods, starting from the captioning approaches that prepared the transition to current cutting-edge visual question answering systems. Recent developments in the field encourage the development of AI systems that will support them by design. Explainability of the decision-making process of AI systems, either built-in or as a by-product of the acquired reasoning capabilities, underpins the understanding of those systems robustness, their underlying logic, and their improvement potential.

Piotr Borkowski, Krzysztof Ciesielski, and Mieczysław A. Kłopotek base their work on the known phenomenon that text document classifiers may benefit from the inclusion of hypernyms of the terms in the document. The authors have elaborated a new type of document classifiers, so-called semantic classifiers, trained not on the original data but rather on the categories assigned to the document by our semantic categorization that requires significantly smaller corpus of training data and outperforms traditional classifiers used in the domain. With this research, the authors want to clarify what is the advantage/disadvantage of using supercategories of the assigned categories (an analogon of hypernyms) on the quality of classification.

Damian Węgrzyn, Piotr Wrzeciono, and Alicja Wieczorkowska present the usage of deep learning in flue pipe-type recognition. Organbuilders claim that they can distinguish the pipe mouth type only by hearing it, and the authors used artificial neural networks (ANN) to verify if it is possible to train ANN to recognize the details of the organ pipe, as this confirms a possibility that a human sense of hearing may be trained as well. In the future, usage of deep learning in the recognition of pipe sound parameters may be used in the voicing of the pipe organ and the selection of appropriate parameters of pipes to obtain the desired timbre.

In the following group of papers, different perspectives on applicability of scientific work in industrial settings are presented. Lothar Hotz, Rainer Herzog, and Stephanie von Riegen address challenges in mechanical and plant engineering, specifically those related to the adaption to changing requirements or operating conditions at the plant operator's premises. Such changes require a well-coordinated cooperation with the machine manufacturer and his suppliers and involve high efforts due to the communication and delivery channels. An autonomous acting machine would facilitate this process. In the paper, subtasks for the design of autonomous adaptive machines are identified and discussed.

Cristian Vidal-Silva, José Ángel Galindo, Jesús Giráldez-Cru, and David Benavides have identified the problem that the completion of partial configurations represents an expensive computational task. Existing solutions, such as those which use modern constraint satisfaction solvers, perform a complete search, making them unsuitable on large-scale configurations. In their work, the authors propose an approach based on diagnosis tasks based on an algorithm named FastDiag, an efficient solution for preferred minimal diagnosis (updates) in the context of partial configuration.

Chiara Grosso and Cipriano Forza emphasize the increasing demand for online transactions. This is propelled by both the digital transformation paradigm and the COVID-19 pandemic. The research on web infrastructure design recognizes the impact that social, behavioral, and human aspects have on online transactions in e-commerce, e-health, e-education, and e-work. The authors present a study focusing on the social dimension of the e-commerce of customizable products. This domain was selected because of the specificity of its product self-design process in terms of customers' decision-making and their involvement in product value creation. The results should provide companies and software designers with insights about customers' need for social presence during their product self-design experience.

In their paper, Ignacio Romero, Jorge Estrada, Angel L. Garrido, and Eduardo Mena point out that traditional media are experiencing a strong change. The collapse of advertising-based revenues on paper newspapers has forced publishers to concentrate efforts on optimizing the results of online newspapers published on the web by improving content management systems. The authors present an approach for performing automatic recommendation form news in this hard context combining matrix factoring and semantic techniques. The authors have implemented their solution in a modular architecture design to give flexibility to the creation of elements that take advantage of these recommendations and also with great monitoring possibilities. Experimental results in real environments are promising, improving outcomes regarding traffic redirection and clicks on ads.

The work of Viet-Man Le, Thi Ngoc Trang Tran, and Alexander Felfernig investigates feature model-based configuration which involves selecting desired features from a collection of features (called a feature model) that satisfy pre-defined constraints. Configurator development can be performed by different stakeholders with distinct skills and interests, who could also be non-IT domain experts with limited technical understanding and programming experience. In this context, a simple configuration framework is required to facilitate non-IT stakeholders' participation in configurator development processes. In their paper, the authors present a tool named FM2EXCONF that enables stakeholders to represent configuration knowledge as an executable representation in Microsoft Excel. The tool supports the conversion of a feature model into an Excel-based configurator, which is performed in two steps. In the first step, the tool checks the consistency and anomalies of a feature model. As a second feature, explanations (which are included in the Excel-based configurator) are provided to help non-IT stakeholders to fix inconsistencies in the configuration phase.

The last two papers in the track are emphasizing different aspects of basic research and algorithmic problems in the field. The paper of Antoni Ligęza, Paweł Jemioło, Weronika T. Adrian, Mateusz Ślażyński, Marek Adrian, Krystian Jobczyk, Krzysztof Kluza, Bernadetta Stachura-Terlecka, and Piotr Wiśniewski explores a "yet another approach" to explainable artificial intelligence. The proposal consists in the application of constraint programming to discover internal structure and parameters of a given black-box system. Apart from specification of a sample of the input and output values, some presupposed knowledge about the possible internal structure and functional components is required. This knowledge can be parameterized with respect to the functional specification of internal components, connections among them, and internal parameters. Models of constraints are put forward, and example case studies illustrate the proposed ideas.

Frej Berglind, Jianhua Chen, and Alexandros Sopasakis compare classic scalar temporal difference learning with three new distributional algorithms for playing the game of five-in-a-row using deep neural networks based on the application of different algorithms. The algorithms are applicable to any two-player deterministic zero sum game. Though all the algorithms utilized performed reasonably well, some advantages and disadvantages were identified which are emphasized in the paper.

Despite the difficulties related to COVID-19 which hindered us (the organizers) to carry out the ISMIS conference and the industrial track physically in the area of Graz, the papers presented were of very high quality, presented in the form pre-recorded videos and complemented with live discussion sections. It is a great pleasure to thank all the people who helped this book come into being and made ISMIS 2020 in general and the industrial track in particular a successful and exciting event. We would like to express our appreciation for the work of the ISMIS 2020 program committee members and external reviewers who helped assure the high standard of accepted papers. We would like to thank all authors, without whose high-quality contributions, it would not have been possible to organize the conference. We are grateful to all the organizers and contributors to the successful preparation and implementation of ISMIS 2020. We are thankful to the people at Springer for supporting the ISMIS 2020 and the possibility to publish this extra volume for the industrial section. We believe that this book will become a valuable source of reference for your ongoing and future research activities.

Martin Stettinger
Gerhard Leitner
Alexander Felfernig
Zbigniew W. Ras

Organization

Editors

Martin Stettinger	Graz University of Technology
Gerhard Leitner	University of Klagenfurt
Alexander Felfernig	Graz University of Technology
Zbigniew Ras	University of North Carolina

Program Committee

Esra Akbas	Oklahoma State University
Marharyta Aleksandrova	University of Luxembourg
Aijun An	York University
Troels Andreasen	Roskilde University
Annalisa Appice	University Aldo Moro of Bari
Martin Atzmueller	Tilburg University
Arunkumar Bagavathi	Oklahoma State University
Ladjel Bellatreche	Poitiers University
Robert Bembenik	Warsaw University of Technology
Petr Berka	University of Economics, Prague
Maria Bielikova	Slovak University of Technology in Bratislava
Gloria Bordogna	National Research Council of Italy-CNR
Jose Borges	University of Porto
François Bry	Ludwig Maximilian University of Munich
Jerzy Błaszczyński	Poznań University of Technology
Michelangelo Ceci	Universita degli Studi di Bari
Jianhua Chen	Louisiana State University
Silvia Chiusano	Politecnico di Torino
Roberto Corizzo	UNIBA
Alfredo Cuzzocrea	ICAR-CNR and University of Calabria
Marcilio De Souto	LIFO/University of Orleans

Luigi Di Caro	University of Torino
Stephan Doerfel	Micromata
Peter Dolog	Aalborg University
Dejing Dou	University of Oregon
Saso Dzeroski	Jozef Stefan Institute
Christoph F. Eick	University of Houston
Tapio Elomaa	Tampere University of Technology
Andreas Falkner	Siemens AG Österreich
Nicola Fanizzi	Università degli studi di Bari "Aldo Moro"
Stefano Ferilli	Universita' di Bari
Gerhard Friedrich	Alpen-Adria-Universitat Klagenfurt
Naoki Fukuta	Shizuoka University
Maria Ganzha	Warsaw University of Technology
Paolo Garza	Politecnico di Torino
Martin Gebser	University of Klagenfurt
Bernhard Geiger	Know-Center GmbH
Michael Granitzer	University of Passau
Jacek Grekow	Bialystok Technical University
Mohand-Said Hacid	Université Claude Bernard Lyon 1-UCBL
Hakim Hacid	Zayed University
Allel Hadjali	LIAS/ENSMA
Mirsad Hadzikadic	UNC Charlotte
Ayman Hajja	College of Charleston
Alois Haselboeck	Siemens AG
Shoji Hirano	Shimane University
Jaakko Hollmén	Aalto University
Andreas Holzinger	Medical University and Graz University of Technology
Andreas Hotho	University of Wuerzburg
Lothar Hotz	University of Hamburg
Dietmar Jannach	University of Klagenfurt
Adam Jatowt	Kyoto University
Roman Kern	Know-Center GmbH
Matthias Klusch	DFKI
Dragi Kocev	Jozef Stefan Institute
Roxane Koitz	Graz University of Technology
Bozena Kostek	Gdansk University of Technology
Mieczysław Kłopotek	Polish Academy of Sciences
Dominique Laurent	Université Cergy-Pontoise
Marie-Jeanne Lesot	LIP6 - UPMC
Rory Lewis	University of Colorado at Colorado Springs
Elisabeth Lex	Graz University of Technology
Antoni Ligeza	AGH University of Science and Technology
Yang Liu	Hong Kong Baptist University
Jiming Liu	Hong Kong Baptist University

Corrado Loglisci	University of Bari
Henrique Lopes	Cardoso University of Porto
Donato Malerba	Università degli Studi di Bari "Aldo Moro"
Giuseppe Manco	ICAR-CNR
Yannis Manolopoulos	Open University of Cyprus
Małgorzata Marciniak	Institute of Computer Science PAS
Mamoun Mardini	University of Florida
Elio Masciari	Federico II University
Paola Mello	University of Bologna
João Mendes-Moreira	University of Porto
Luis Moreira-Matias	NEC Laboratories Europe
Mikolaj Morzy	Poznan University of Technology
Agnieszka Mykowiecka	IPI PAN
Tomi Männistö	University of Helsinki
Mirco Nanni	ISTI-CNR Pisa
Amedeo Napoli	LORIA Nancy (CNRS-Inria-Université de Lorraine)
Pance Panov	Jozef Stefan Institute
Jan Paralic	Technical University Kosice
Ruggero G. Pensa	University of Torino, Italy
Jean-Marc Petit	Université de Lyon, INSA Lyon
Ingo Pill	Graz University of Technology
Luca Piovesan	DISIT, Università del Piemonte Orientale
Olivier Pivert	IRISA-ENSSAT
Lubos Popelinsky	Masaryk University
Jan Rauch	University of Economics, Prague
Marek Reformat	University of Alberta
Henryk Rybiński	Warsaw University of Technology
Hiroshi Sakai	Kyushu Institute of Technology
Tiago Santos	Graz University of Technology
Christoph Schommer	University of Luxembourg
Marian Scuturici	LIRIS-INSA de Lyon, France
Nazha Selmaoui-Folcher	University of New Caledonia
Giovanni Semeraro	University of Bari
Samira Shaikh	UNC Charlotte
Dominik Slezak	University of Warsaw
Urszula Stanczyk	Silesian University of Technology
Jerzy Stefanowski	Poznan University of Technology, Poland
Marcin Sydow	PJIIT and ICS PAS, Warsaw
Katarzyna Tarnowska	San Jose State University
Herna Viktor	University of Ottawa
Simon Walk	Graz University of Technology

Alicja Wieczorkowska	Polish-Japanese Academy of Information Technology
David Wilson	UNC Charlotte
Yiyu Yao	University of Regina
Jure Zabkar	University of Ljubljana
Slawomir Zadrozny	Systems Research Institute, Polish Academy of Sciences
Wlodek Zadrozny	UNC Charlotte
Bernard Zenko	Jozef Stefan Institute
Beata Zielosko	University of Silesia
Arkaitz Zubiaga	Queen Mary University of London

Additional Reviewers

Max Toller
Henryk Rybiński
Allel Hadjali
Giuseppe Manco
Aijun An
Michelangelo Ceci
Giovanni Semeraro
Michael Granitzer
Simon Walk

Contents

Industrial Applications

Basic Research and Algorithmic Problems

Applications in the Automotive and Transport Sector

Parameter Tuning for Speed Changes Detection in On-Road Audio Recordings of Single Drives

Elżbieta Kubera[1(✉)], Alicja Wieczorkowska[2], and Andrzej Kuranc[1]

[1] University of Life Sciences in Lublin, Akademicka 13, 20-950 Lublin, Poland
{elzbieta.kubera,andrzej.kuranc}@up.lublin.pl
[2] Polish-Japanese Academy of Information Technology, Koszykowa 86, 02-008 Warsaw, Poland
alicja@poljap.edu.pl

Abstract. Economical driving not only saves fuel, but also reduces the carbon dioxide emissions from cars. Apart from environmental benefits, road safety is also increased when drivers avoid speeding and sudden changes of speeds. However, speed measurements usually do not reflect speed changes. In this paper, we address automatic detection of speed changes, based on audio on-road recordings, which can be taken at night and at low-vision conditions. In our approach, the extraction of information on speed changes is based on spectrogram data, converted to black-and-white representation. Next, the parameters of lines reflecting speed changes are extracted, and these parameters become a basis for distinguishing between three classes: accelerating, decelerating, and maintaining stable speed. Theoretical discussion of the thresholds for these classes are followed by experiments with automatic search for these thresholds. In this paper, we also discuss how the choice of the representation model parameters influences the correctness of classification of the audio data into one of three classes, i.e. acceleration, deceleration, and stable speed. Moreover, for 12-element feature vector we achieved accuracy comparable with the accuracy achieved for 575-element feature vector, applied in our previous work. The obtained results are presented in the paper.

Keywords: Driver behavior · Hough transform · Intelligent transportation systems

1 Introduction

Measurements of vehicle speed on public roads have been occupying the minds of scientists in various fields of science, economy and social life for a long time. Extensive research has been done in the field of road safety, because excessive speed is indicated as the cause of numerous road accidents [1,2]. Moreover, many

Partially supported by research funds sponsored by the Ministry of Science and Higher Education in Poland.

M. Stettinger et al. (Eds.): ISMIS 2020, SCI 949, pp. 3–14, 2021.
https://doi.org/10.1007/978-3-030-67148-8_1

studies related to vehicle speed measurements also discuss and investigate the environmental impact of vehicles [3–6], both in urban driving conditions, as well as in motorway traffic [7,8]. Growing deterioration of air quality in urban agglomerations is largely associated with the increase in the number of road transport means and their deteriorating technical condition. The problem exacerbates when climatic conditions hinder spontaneous purification of the air in strongly urbanized areas. Therefore, various actions are undertaken to make vehicular traffic more fluent and optimized in terms of traffic safety, fuel consumption and emissions of harmful exhaust components [9]. Efforts to influence travel behavior in support of reducing emissions and congestion have been undertaken since the 1970s [10].

Intelligent vehicle traffic monitoring and controlling systems optimize traffic through speed measurement and the classification of vehicles [11–14]. Transport agencies often use speed measurements as the basis of decisions such as setting speed limits, synchronizing traffic signals, placing road signs and then determining the effectiveness of the steps taken [15].

Another problem is to assess whether the observed speed change reflects the driver's intention to accelerate or decelerate, or this change is negligible, and the driver's intention was to maintain approximately constant speed. We discuss further in this paper what speed changes can be considered intentional or not. The experiments with automatic classification of speed changes may serve as a tool of verification if the discussed thresholds of speed changes for discerning stable speed and intentional deceleration/acceleration work well as a classification criterion.

It should be noted that excessive speed and sudden speed changes cause many accidents; this has been confirmed in the detailed studies on road events, their causes and consequences [1,2,16,17]. According to [18], the greater the speed variability, the greater interaction between vehicles in traffic and the associated danger. Moreover, it should be emphasized that the greater the speed variability, the greater vehicle energy demand, the higher fuel consumption, and the higher emissions [19]. Dynamic, unsteady load states of internal combustion engines during acceleration are associated with the occurrence of imperfections in the fuel combustion process, and they implicate increased emission of toxic exhaust components, including particulate matter. The frequent acceleration combined with frequent and intense deceleration of the vehicle results in an increased emission of dust from the brake linings in the brake mechanisms and from rubber friction products formed due to wear of vehicle tires [5,20].

The optimized traffic, without congestion and unjustified changes in speed, results in the least onerous impact of vehicles on the environment and is relatively safe. These issues are analyzed around the world and are of interest of governments, due to the serious consequences they have for human health [21,22]. The vehicle speed monitoring is therefore an important aspect in tackling harmful emissions, and it provides the necessary information for public administration (e.g. European Environment Agency) to improve the transport management [23].

Speed measurements are the basis for modeling the vehicle traffic and its impact [24]. However, some driver behaviors are difficult to investigate and require long-term observation, for example the analysis of the vehicular traffic near speed measuring points. Such an analysis can easily discover drivers who usually exceed the speed limits, then reduce their speeds momentarily only near enforcement locations, and next accelerate again. This behavior (called kangaroo effect) is dangerous, and it also contributes to excessive emissions.

Acoustic methods can be used to classify vehicles and assess changes in their speed, see [25–27]. The obtained results indicate the great potential of these techniques and the possibility of supplementing currently used methods of measuring the speed with the measurement of the acceleration of the vehicles.

2 Detecting Speed Changes from Audio Data

There exist many techniques for speed measurements, including Doppler radar, video image-based detection, and using various sensors (infra-red, and also acoustic sensors). Average speed measurements are also taken. However, to the best of our knowledge, no other researchers worked on automatic speed change detection, except our team [26–28]. We use audio data as a basis, as they can be obtained at night and at low visibility conditions. Spectrogram for an audio recording of a single car approaching the recorder, then passing by, and driving away is shown in Fig. 1. We can observe lines before and after passing the microphone, whereas the central part shows curves, as this part is heavily affected by the Doppler effect. These lines correspond to speed changes of the recorded car.

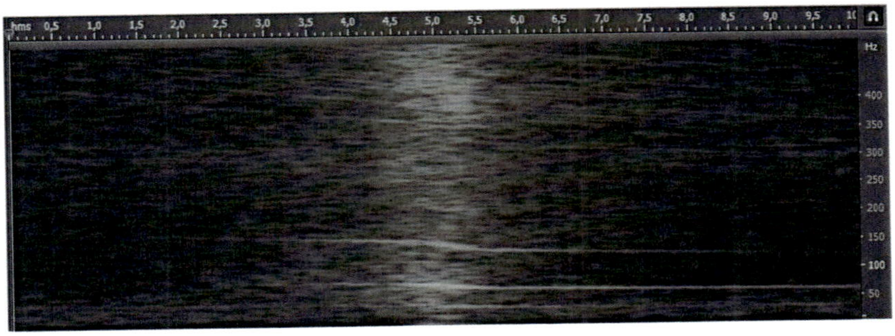

Fig. 1. Grayscale spectrogram for a single channel of audio data (for deceleration). The moment of passing the recorder is in the middle of the graph. The graph illustrates changes of frequency contents over time. Higher brightness corresponds to higher level. Fast Fourier Transform (FFT) was used to calculate spectra in consecutive time frames

In our previous works, we detected speed changes from both test-bed and on-road recordings. Ten-second audio segments, centered at the moment of passing the microphone, were used in these experiments. We aimed at recognizing one of 3 classes: acceleration, stable speed, and deceleration. For on-road recordings, we obtained 99% accuracy for 84 drives of a single car, with 28 drives per class. In tests on data representing 3 other cars, 75% was obtained. Next, we prepared a set of recordings for 6 cars, recorded in 3 seasons: winter, spring, and summer. For these data, we obtained 92.6% for a 109-element feature set, and 94.7% for 575 features [29]. When we applied image-based approach, with grayscale spectrogram transformed to binary (black-and-white) images, we obtained almost 80% for a single feature. The main idea behind these works was to extract lines from spectrograms. This task poses a lot of difficulties, as there is a lot of noise in spectrograms, and the lines are curved at the moment of passing the microphone (where the energy is the highest). Still, we can observe that the slope of lines corresponds to the speed changes: sloping down for deceleration (see Fig. 1), being almost horizontal for stable speed, and going up for acceleration, except the moment of passing the microphone.

The problems we have to solve in this approach include also grayscale to binary image conversion, and selection of border slopes for each class. Hough transform has been applied to line detection, taking binary images as input [30]. Solving these problems is the goal of our paper.

3 Methodology

In this work, we address the issues related to thresholds selection in grayscale-to-binary image conversion, and in edge detection, for the purpose of detecting lines corresponding to speed changes in spectrograms. We also address selecting the limits of slopes/speeds for each class. The grayscale to binary conversion is performed using two approaches: threshold-based conversion, and Canny edge detection (which requires selecting 2 thresholds) [31].

3.1 Audio Data

The audio data we used in this work represent on-road recordings, acquired using Mc Crypt DR3 Linear PCM Recorder, with 2 integrated high-quality microphones (48 kHz/24 bit, stereo). 318 drives were recorded, each one representing one of our 3 target classes: 113 for deceleration, 94 for stable speed, and 111 for acceleration. Each drive represents one car only (of 6 cars used).

In our previous work we used 10 s audio segments, namely 5 s for approaching the microphone and 5 s after passing it. However, we observed that such a segment is too long, and the slopes of lines in the spectrogram may change in this segment. Therefore, we decided to analyze 3 s long segments, more appropriate for 60 m long road segment and the speed range used, in order to obtain approximately constant acceleration or deceleration values. The spectrum range was limited to 300 Hz.

Hough Transform for Line Detection. The output of the Hough technique indicates the contribution of each point in the image to the physical line. Line segments are expressed using normals: $x\cos(\theta) + y\sin(\theta) = r$, where $r \geq 0$ is the length of a normal, measured from the origin to the line, and θ is the orientation of the normal wrt. the x axis; x, y - image point coordinates. The plot of the possible r, θ values, defined by each point of line segments, represents mapping to sinusoids in the Hough parameter space. The transform is implemented by quantizing the Hough parameter space into accumulator cells, incremented for each point which lies along the curve represented by a particular r, θ. Resulting peaks in the accumulator array correspond to lines in the image. The more points on the line (even discontinuous), the higher the accumulator value, so the maximum corresponds to the longest line. For $\theta = 0$ [°] the normal is horizontal, so the corresponding line is vertical, and $\theta = 90°$ corresponds to horizontal line; $r > 0$ is expressed in pixels. We limit our search to [45°, 135°], which covers lines of interest for us, i.e. horizontal and sloping a bit.

Feature Vector. We use a very simple representation of spectrograms, namely the maximum of the accumulator and its corresponding θ and r for each 3 s segment of the spectrogram, i.e. detecting the longest line in this segment, for each channel of audio data. As a result, we have 12 features for each drive, i.e. for 3 s of approaching the microphone and 3 s after passing the microphone, for both left and right channel of the audio data.

3.2 Thresholds

In our previous work, we also dealt with selecting thresholds for grayscale to binary image conversion, and in the Canny algorithm, before applying Hough transform [28]. We compared visually 7 versions of thresholds, adaptive and fixed (uniform), with arbitrarily chosen fixed values. In adaptive thresholding, the thresholds are changed locally, i.e. depending on the local luminance level. The mean and the gaussian-weighted sum of neighboring values were tested, minus constant c = 2. In uniform thresholding, pixels are set to white if their luminance level is above a predefined level, otherwise they are set to black. Image normalization was performed as preprocessing, so the luminance in our grayscale spectrograms was within [0, 255]. Fixed thresholding with threshold equal to 80% of the highest luminance yielded the best results. In the Canny edge detection applied as preprocessing before Hough transform, the pixel is accepted as an edge, if its gradient is higher than the upper threshold, and rejected if its gradient is below the lower threshold. Thus, 2 thresholds are needed. The spectrum was limited to [10, 300] Hz in this case.

The parameter space was not thoroughly searched in our previous work, as we had too many options to check. In this paper, we decided to address threshold tuning. Since fixed thresholds worked best in our previous work, we decided to test several versions of fixed thresholds, namely from 70% to 95% of luminance applied as criterion to assign black or white. In the Canny algorithm, the proportions of thresholds between 2:1 and 3:1 are advised [31], so in this paper we decided to check such pairs, namely {30%, 60%} of the luminance, {30%, 75%}, {30%, 90%}, {40%, 80%}, {40%, 90%}, and {45%, 90%} of the luminance. We also tested another 2 pairs, namely 33% below and above median value of the luminance, as well as 33% below and above mean value of the luminance.

3.3 Limits for Speed Changes

We can assume that acceleration above $0.3\,\mathrm{m/s^2}$, i.e. about 5.4 km/h in 5 s, is an intentional action. We can also assume that deceleration of $-0.25\,\mathrm{m/s^2}$ for 50 km/h speed is intentional (for higher speed, e.g. 140 km/h, greater decrease would be considered as intentional). Also, changes within $[-0.2,\ 0.2]\,\mathrm{m/s^2}$ can be considered unintentional, and if they happen, then the driver is probably intending to maintain constant speed. These changes can be seen as slopes of lines visible in spectrogram, except the Doppler effect, most pronounced at the moment of passing the microphone. The values indicated above correspond to $\pm 2^\circ$ of the slope of the line in the spectrogram, i.e. 88° and 92° for the normal. This discussion shows the proposed limits for classifying speed changes as intentional or not, based on calculation.

3.4 Classification

Since we have a small, 12-element feature set, we decided to apply simple classification algorithms: decision trees and random forests (RF). RF are ensemble classifiers consisting of many decision trees, constructed in a way that reduces the correlation between the trees. Decision tree classifier J4.8 from WEKA (implemented in Java) was applied [32], and RF implementation in R was used in our experiments [33]. J4.8 is a commonly used decision tree classifier.

CV-10 cross-validation was used, calculated 10 times. Additionally, we constructed the following heuristic rule to classify the investigated automotive audio data into acceleration, deceleration, and stable speed classes. Namely, we take θ corresponding to the maximum accumulator among the 4 spectrogram parts for this sound. If $\theta > AccSlope$, the data are classified as acceleration, if $\theta < DecSlope$, then as deceleration, otherwise as stable speed. The thresholds $AccSlope$ and $DecSlope$ were used in 2 versions.

- In the 1st version, they were experimentally found in brute-force search. Since the output of the Hough transform represents the slope of the detected line, in degrees, in integer values, we tested the limit values for classifying lines as acceleration, stable speed, or deceleration, in one-degree-step search.
- In the 2nd version, the limits $[88^\circ, 92^\circ]$ of unintentional speed changes (see Sect. 3.3) were tested.

These rules were tested once on the entire data set.

Additionally, we constructed a decision tree for θ and r corresponding to the maximum of the accumulator (thus actually selecting one of 4 parts of the analyzed spectrogram, where the longest line was found), to obtain an illustrative classification rule. The conditions in the nodes of the tree indicate the boundary values at each step of this commonly used classification algorithm, and reflect the best *AccSlope* and *DecSlope* values for the lines found.

4 Experiments and Results

The results of our experiments are shown in Fig. 2. We would like to emphasize that these results were obtained for up to 12 features, whereas in our previous work we had 575 features [29]. As we can see, the best results were obtained for random forests, especially fixed thresholds in grayscale-to-binary image conversion. The best results were achieved for 95% of maximum luminance (after normalization) threshold, yielding 93.87%, very close to the best result we achieved so far for this set of recordings. Acceleration was never recognized as deceleration in this case, and deceleration was recognized as acceleration in 2 out of 1130 cases.

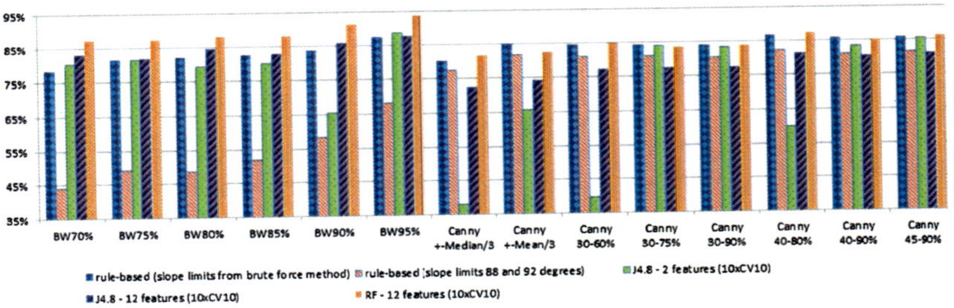

Fig. 2. The results obtained for various thresholds and classification methods. BW indicates fixed thresholds used in grayscale-to-binary (i.e. black and white) image conversion. Percentage values on the horizontal axis indicate thresholds tested. Rule-based classifiers correspond to the 2 versions described in Sect. 3.4

Generally, random forests performed best for fixed thresholds in grayscale-to-binary conversion, whereas Canny algorithm worked well with other classifiers as well, namely with rule-based approach with slope limits found via brute-force search, and sometimes also with decision tree classifiers.

We can also observe that the values 88° and 92°, corresponding to the indications of the limits for intended stable speed (Sect. 3.3), do not work well. They indicate stable speed, but the limits for acceleration and deceleration might be different. The limit values for θ yielding the best results for particular thresholds, found in our brute-force (with 1-degree step) threshold search, are shown in

Fig. 3. As we can see, for the Canny method the limit values are approximately symmetrical wrt. $\theta = 90°$, corresponding to the horizontal line. For uniform thresholding with fixed threshold however, both limit values are always below $\theta = 90°$. This might be caused by the bending, related to the Doppler effect, (see the middle part of Fig. 1), where the lines/curves are most pronounced. Canny method detects the edges of lines, not lines themselves, and these edges might be lost in the noisy background at the moment of passing the mic. When lines (not just their edges) are detected using uniform fixed thresholds, the slope of lines is influenced by the bending at the moment of passing the microphone, i.e. the end of line for the first part of the spectrogram and the beginning for the second part of the spectrogram.

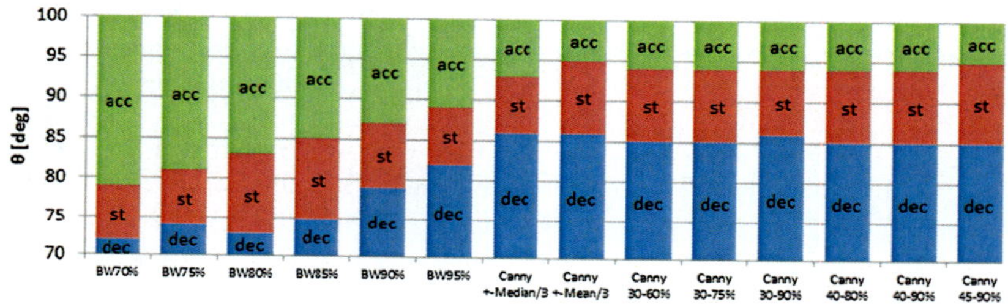

Fig. 3. The limit values for θ, yielding the best results for particular thresholds (dec - deceleration, st - stable speed, acc - acceleration)

In our previous work, we used 5-s segments of spectrograms before and after passing the microphone, as opposed to 3-s segments used here. In the work reported in [28], we obtained the best results for the fixed threshold of 80% of luminance in grayscale-to-binary image conversion, and 12-element feature vector. 80% accuracy was obtained for the decision tree, and 85% for the random forest classifier. Rule-based classification yielded 79% accuracy, when only θ from the Hough transform was applied as a basis of classification. As we can see in Fig. 2, here we obtained 84.6% accuracy for the decision tree, and 88% for the random forest classifier, when 3-s segments of spectrograms were used. Rule-based approach with thresholds found via brute force search yielded 82% in our experiments reported here.

Figure 4 shows the decision tree obtained for the fixed threshold of 95% of luminance in grayscale-to-binary image conversion of the spectrogram image (for the entire data set). As we can see, for $\theta \leq 82°$ acceleration is never indicated in the labels of the left subtree. Also, the limit values are the same as found in our one-degree-step search, which indicated θ pairs (82, 89) and (81, 89) as $(DecSlope, AccSlope)$ yielding the best result. For comparison, Fig. 5 shows the decision tree obtained for grayscale-to-binary image conversion using the Canny method of edge detection with 45% and 90% of luminance as thresholds. As we

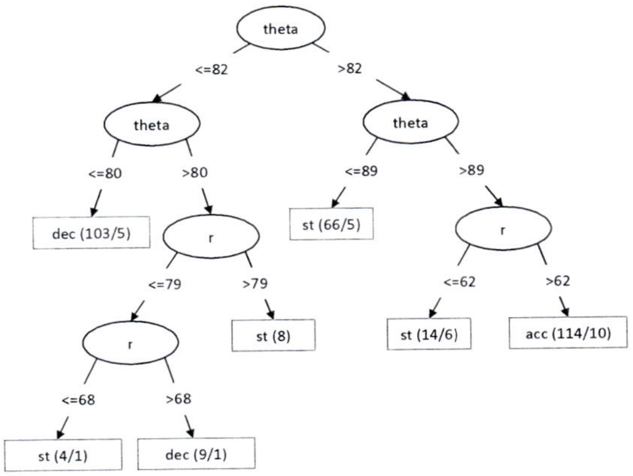

Fig. 4. Decision tree obtained for the fixed threshold of 95% of luminance in grayscale-to-binary image conversion of the spectrogram; this threshold yielded the best results

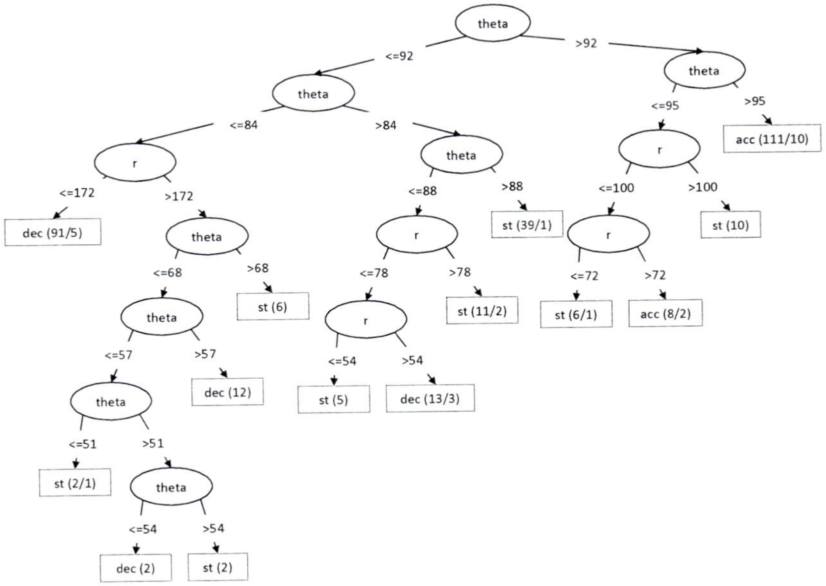

Fig. 5. Decision tree obtained for the Canny method with 45% and 90% thresholds

can see, the conditions in the top nodes indicate the same boundary values as indicated in Sect. 3.3 as limits of no intentional speed changes.

5 Summary and Conclusions

In this paper we present the search for threshold values in grayscale-to-binary spectrogram image conversion for audio data representing on-road recordings of single drives of cars, which either accelerate, decelerate, or maintain stable speed during the recording. The lines that can be seen in the spectrogram correspond to speed changes of the recorded vehicle. We investigated the limit values of the slopes of these lines, both theoretically and in the brute-force threshold search. The limit values obtained in this search coincide partly with the values calculated theoretically as limits of intentional stable speed. We limited the analyzed audio segment to 3 s (as compared to 5 s in our previous experiments), both before and after passing the microphone. In CV-10 crossvalidation repeated 10 times for the random forest classifier and grayscale-to-binary conversion with 95% of luminance threshold, acceleration was never recognized as deceleration, and deceleration was recognized as acceleration only twice. We obtained about 94% accuracy for fixed thresholding in grayscale-to-binary conversion, for 12 features, which is comparable with the best result we achieved so far for 575 features [29]. Therefore, we conclude that this approach is promising, and we are planning to improve this methodology in further works.

References

1. Elvik, R., Vaa, T. (eds.): The Handbook of Road Safety Measures. Emerald Group Pub Ltd., Bingley (2004)
2. Huvarinen, Y., Svatkova, E., Oleshchenko, E., Pushchina, S.: Road safety audit. Transp. Res. Procedia **20**, 236–241 (2017)
3. Czechowski, P.O., Oniszczuk-Jastrząbek, A., Czuba, T.: Eco-driving: behavioural pattern change in Polish passenger vehicle drivers. E3S Web Conf. **28**, 01009 (2018)
4. Fontaras, G., Zacharof, N.-G., Biagio, C.: Fuel consumption and CO_2 emissions from passenger cars in Europe Laboratory versus real-world emissions. Progress Energy Combust. Sci. **60**, 97–131 (2017)
5. Franco, V., Kousoulidou, M., Muntean, M., Ntziachristos, L., Hausberger, S., Dilara, P.: Road vehicle emission factors development: a review. Atmos. Environ. **70**, 84–97 (2013)
6. Sarkan, B., Stopka, O., Gnap, J., Caban, J.: Investigation of exhaust emissions of vehicles with the spark ignition engine within emission control. Procedia Eng. **187**, 775–782 (2017)
7. Bas, E., Tekalp, M., Salman, F.S.: Automatic vehicle counting from video for traffic flow analysis. In: 2007 IEEE Intelligent Vehicles Symposium, Istanbul, Turkey, pp. 392–397. IEEE (2007)
8. Liang, M., Huang, X., Chen, C.-H., Chen, X., Tokuta, A.: Counting and classification of highway vehicles by regression analysis. IEEE Trans. Intell. Transp. Syst. **16**, 2878–2888 (2015)
9. Wang, H., Fu, L., Zhou, Y., Li, H.: Modelling of the fuel consumption for passenger cars regarding driving characteristics. Transp. Res. D **13**(7), 479–482 (2008)
10. Cambridge Systematics: Effects of Travel Reduction and Efficient Driving on Transportation: Energy Use and Greenhouse Gas Emissions. https://www.nrel.gov/docs/fy13osti/55635.pdf. Accessed 20 Dec 2019

11. Adnan, M.A., Sulaiman, N., Zainuddin, N.I., Besar, T.B.H.T.: Vehicle speed measurement technique using various speed detection instrumentation. In: 2013 IEEE Business Engineering and Industrial Applications Colloquium BEIAC, Langkawi, Malaysia, pp. 668–672. IEEE (2013)
12. Balid, W., Tafish, H., Refai, H.H.: Intelligent vehicle counting and classification sensor for real-time traffic surveillance. IEEE Trans. Intell. Transp. Syst. **19**(6), 1784–1794 (2018)
13. Meiring, G.A.M., Myburgh, H.C.: A review of intelligent driving style analysis systems and related artificial intelligence algorithms. Sensors **15**(12), 30653–30682 (2015)
14. Wang, W., Xi, J., Chong, A., Li, L.: Driving style classification using a semisupervised support vector machine. IEEE Trans. Hum.-Mach. Syst. **47**, 650–660 (2017)
15. Schroeder, B.J., Cunningham, C.M., Findley, D.J., Hummer, J.E., Foyle, R.S.: Manual of Transportation Engineering Studies, 2nd edn. Institute of Transportation Engineers, Washington, DC (2010)
16. Banovic, N., Buzali, T., Chevalier, F., Mankoff, J., Dey, A.: Modeling and understanding human routine behavior. In: ACM CHI Conference on Human Factors in Computing Systems 2016, Santa Clara, CA, pp. 248–260. ACM (2016)
17. Bonsall, P.W., Liu, R., Young, W.: Modelling safety-related driving behaviour - the impact of parameter values. Transp. Res. A **39**(5), 425–444 (2005)
18. Mehar, A., Chandra, S., Velmurugan, S.: Speed and acceleration characteristics of different types of vehicles on multi-lane highways. Eur. Transp. **55**(1), 1–12 (2013)
19. Ahn, K., Rakha, H., Trani, A., Van Aerde, M.: Estimating vehicle fuel consumption and emissions based on instantaneous speed and acceleration levels. J. Transp. Eng. **128**(2), 182–190 (2002)
20. Kousoulidou, M., Ntziachristos, L., Mellios, G., Samaras, Z.: Road-transport emission projections to 2020 in European urban environments. Atmos. Environ. **42**(32), 7465–7475 (2008)
21. Health Effects Institute: Diesel exhaust: critical analysis of emissions, exposure, and health effects. Health Effects Institute, Boston, MA (1995)
22. Kim, K.H., Kabir, E., Kabir, S.: A review on the human health impact of airborne particulate matter. Environ. Int. **74**, 136–143 (2015)
23. Ntziachristos, L., Samaras, Z. (Lead authors): EMEP/EEA air pollutant emission inventory guidebook 2019. 1.A.3.b.i-iv Road transport 2019. European Environment Agency, Copenhagen, Denmark (2019)
24. Silva, C.M., Farias, T.L., Frey, H.C., Rouphail, N.M.: Evaluation of numerical models for simulation of real-world hot-stabilized fuel consumption and emissions of gasoline light-duty vehicles. Transp. Res. D-Trans. Environ. **11**(5), 377–385 (2006)
25. Kotus, J.: Determination of the vehicles speed using acoustic vector sensor. In: 2018 SPA, Signal Processing: Algorithms, Architectures, Arrangements, and Applications, Poznan, Poland, pp. 64–69 (2018)
26. Kubera, E., Wieczorkowska, A., Słowik, T., Kuranc, A., Skrzypiec, K.: Audio-based speed change classification for vehicles. In: NFMCP 2016. LNAI, vol. 10312, pp. 54–68. Springer, Cham (2017)
27. Wieczorkowska, A., Kubera, E., Koržinek, D., Słowik, T., Kuranc, A.: Time-frequency representations for speed change classification: a pilot study. In: ISMIS 2017. LNAI, vol. 10352, pp. 404–413. Springer, Cham (2017)
28. Kubera, E., Wieczorkowska, A., Kuranc, A.: Hough transform as a tool for the classification of vehicle speed changes in on-road audio recordings. In: Ceci, M., Loglisci, C., Manco, G., Masciari, E., Ras, Z. (eds.) NFMCP 2019, Revised Selected Papers, LNAI, vol. 11948, pp. 137–154. Springer, Cham (2020). https://doi.org/10.1007/978-3-030-48861-1_9

29. Kubera, E., Wieczorkowska, A., Kuranc, A., Słowik, T.: Discovering speed changes of vehicles from audio data. Sensors **19**(14), 3067 (2019)
30. Fisher, R., Perkins, S., Walker, A., Wolfart, E.: Hypermedia Image Processing Reference. Wiley, West Sussex (2000)
31. Canny, J.: A computational approach to edge detection. IEEE Trans. Pattern Anal. Mach. Intell. **8**(6), 679–98 (1986)
32. Frank, E., Hall, M.A, Witten, I.H.: The WEKA Workbench. Online Appendix for "Data Mining: Practical Machine Learning Tools and Techniques", 4th edn. Morgan Kaufmann (2016)
33. R Core Team: R: A language and environment for statistical computing. R Foundation for Statistical Computing, Vienna, Austria (2017). https://www.R-project.org/

Attempt to Better Trust Classification Models: Application to the Ageing of Refrigerated Transport Vehicles

Marie Le Guilly[1(✉)], Claudia Capo[1,2,3], Jean-Marc Petit[1], Vasile-Marian Scuturici[1], Rémi Revellin[2], Jocelyn Bonjour[2], and Gérald Cavalier[3]

[1] Univ Lyon, INSA Lyon, LIRIS, UMR 5205 CNRS, Villeurbanne, France
marie.le-guilly@insa-lyon.fr
[2] Univ Lyon, INSA Lyon, CETHIL, UMR 5008 CNRS, Villeurbanne, France
[3] Cemafroid, Fresnes, France

Abstract. CEMAFROID is a company with a french delegated public service, delivering conformity attestations of refrigerated transport vehicles. It studies the ageing of those vehicles, depending on several physiochemical and mechanical factors, for which physical models have been proposed by researchers in thermal engineering. The DATAFRIG® database records more than 300 000 attestations of vehicles over 80 attributes, opening the opportunity to predict the ageing by building a numerical model using machine learning methods. During the development of such a model, several classical questions appeared, regarding the data quality, the field reality and the mistrust of domain experts. In this paper, we propose to use the notion of functional dependencies to address the aforementioned model's limitations. In particular, we investigate how FDs could help, especially using their counterexamples, that turn out to provide meaningful examples of such limitations, easily interpretable by domain experts. Interestingly, the existence of such counterexamples in the dataset is a way of demystifying the numerical model with the experts, by giving them back the control over their own data. This approach has been tested with domain experts from CEMAFROID, with many positive feedbacks. It is worth noting that this attempt to better trust classification models is not limited to a particular application, and could be generalized to others.

1 Introduction

Refrigerated transport vehicles' main objective is to supply consumers with good quality and safe perishable products. A wide array of services to the cold chain is provided by using a diversified fleet that is continually evolving. In this context, CEMAFROID offers testing and calibration services, and has been designated as an approved body for issuing conformity attestations for refrigerated transport vehicles. To this end, researchers in thermal engineering have access to a large database, DATATFRIG®, recording data about more than 300 000 attestations

M. Stettinger et al. (Eds.): ISMIS 2020, SCI 949, pp. 15–27, 2021.
https://doi.org/10.1007/978-3-030-67148-8_2

for refrigerated vehicles, with tables spanning over almost 80 attributes. Among all these attributes, the thermal insulation of the body is a basic element of the refrigerated equipment, as it is one of the major barriers again heat gain from the outside into the refrigerated enclosure. This attribute is characterized by an "insulation coefficient" denoted K: it is very important as it may be controlled as required by the ATP (the international Agreement for the Transport of Perishable foodstuff). The lower the K coefficient the better, but due to the ageing of the vehicle, it increases over time. To control it and allow the vehicle to continue to transport perishable food, its thermal insulation is measured again after 12 years of service. The ratio between this value and the initial value allows too define the *ageing* of the vehicle, a value that can greatly differ from one vehicle to another.

To explain the variability in the ageing of refrigerated vehicles, several studies have looked at the problem to identify factors to explain it, and propose physical models for the ageing. [7] presented the factors playing a role in the ageing process, classifying them into three different categories. The effects of some of these factors were analyzed in previous studies. As stated by [14], the ageing of an insulated enclosure for a refrigerated vehicle is mostly due to the permeability of the foam to the gases, to the condensation of water into the foam cells and to the increase of the percentage of the broken cells. These causes explain the ageing from a physio-chemical point of view and primarily concern the panels used to build the insulated enclosure and the insulation materials itself. Ageing of refrigerated vehicles also has a mechanical component due to the movements on the road, the routes covered and the payload. A statistical analysis carried out in [13] highlighted the influence in the ageing rate due to the rails and to the refrigerating units. As pointed out by the authors, in presence of these accessories, the mechanical stresses are deeply increased.

But whereas all these studies rely on physical models to analyse the problem, the amount of data available in the DATATFRIG® makes it possible to build a predictive model, using machine learning techniques: such a new approach is a way to tackle the problem from a different angle, to possibly identify new causes for the ageing of vehicles, and to confront new results to the one obtained with physical models. We propose in the present paper to approach the problem from a different angle, by developing a numerical model to predict the ageing of vehicles. Such a numerical model is possible owing to the existence of an important database, that we propose to use in order to build a classification model, relying on well-known algorithm in machine learning. We propose to use the attributes identified in the aforementioned studies as features for the classifier, in order to predict the computed ageing attribute. This classification problem, as well as the data processing and the first classification results are presented in this paper. Shifting from a physical to a numerical model is indeed a big change of paradigm, especially for the domain experts that mainly design and work with physical models. The ML techniques might even generate scepticism. In this study, we therefore paid strong attention to give guaranties about the model, and to explain its limitations. It was important to explain why the considered data could (or not) produce a satisfying model, so that the experts that would use it afterwards could trust it and understand why its decisions made sense. To this end, it appeared that functional dependencies could be a very powerful tool to model knowledge on the classifier [11].

Functional dependencies (FDs) are a well-known and widely studied notion over the years, at the foundation of the theory for relational database design (see [2]), expressing constraints and relationships between two sets of attributes, that are useful to design databases schemas, but can also be used in data cleaning or for query optimization. In the setting of this paper, we propose to use FDs to model the possible limitations of a predictive model, by considering that FDs can determine whether a model can actually exists over a relation, before actually trying to determine the model using a classification algorithm. Indeed, a classification model seeks an optimal function, with respect to a given error measure, that maps features to a class. This process relies on a strong hypothesis, which is the existence of a function from the features to the labels, and only seeks to determine such a function. However, in a complementary way, FDs can be seen as a way to determine if such a function even exists. A detailed study of the link between FDs and classification is available in [11]. We only underline here that when there exists a function from a feature vector X to a label Y, it follows that the function dependency $X \rightarrow Y$ is satisfied.

Thus, for a classification problem, it seems logical to first verify the existence of a function using FD $features \rightarrow class$. Moreover, understanding what tuples prevent the FD from being satisfied can lead to improve the results, if the identified blockages can be explained and fixed by the domain experts. This is where the notion of counterexample is useful. Counterexamples identify pairs of tuples for which the classifier will never be able to perform correctly, as for the same input, it will always predict the same output. Understanding where these counterexamples come from, and what can be done to avoid having such tuples in the dataset, is then an important tool to improve the classification performances.

These notions were used in an application to CEMAFROID's data, for which the notion of counterexamples appeared to be central for the discussion with thermal engineers. From this statement, we retrieve tuples that can impact the performances of a classification algorithms on CEMAFROID's data. In this paper, we expose how we presented such counterexamples to the experts in thermal engineering, in order to give them back the control over their own data, by explaining some of the limitations of the classifier. The discussion that followed allowed to identify several possible causes for the existence of counterexamples, that are exposed in this paper, along with possible solutions to remove them. We also measured the proportion of counterexamples in the dataset, by using metrics on the error of FDs in a relation. The feedback obtained during this discussion highlighted several interesting research problems related to classification, for which FDs could be useful, such as feature engineering and data cleaning.

The contributions of this paper are as follows:

- The presentation of a classification problem for the ageing of refrigerated transport vehicles, along with the corresponding dataset.

- The proposition of a classification model, giving very encouraging accuracy and precision measures.
- The analysis of counterexamples from the dataset with domain experts, and the conclusions to draw from it.
- Metrics to analyze the proportion of counterexamples in the dataset.

Paper Organisation. Section 2 presents the available data and the first numerical model built on it. Section 3 presents the approach used, and the results of a discussion with thermal engineers, based on the analysis of counterexamples to a FD in their dataset, before concluding in Sect. 5.

2 Predicting the Ageing of Refrigerated Transport Vehicles Data

2.1 Classification Problem

In refrigerated transport, the thermal insulation of the body is one of the basic elements of the refrigerated equipment. The efficiency of this thermal insulation is quantified by the body's overall insulation coefficient ("K coefficient"), a parameter increasing during the time due to the ageing of the body's insulation. The evolution of K with time can be very different from one vehicle to another due to many factors, whose analysis and modeling has been previously treated by several studies [4], highlighting various facets of this phenomenon.

The ageing of a refrigerated vehicle used in the carriage of perishable foodstuffs is quantified through the evolution over the time of the K coefficient. The French Regulation requires that this coefficient is measured for all the in-service vehicles after twelve years of service. As a result, the *ageing* evaluation may be made by comparing the K coefficient value of the in-service vehicle at the twelfth year of life, K_{12} to the initial K value of the prototype equipment, K_p, through their ratio:

$$ageing = \frac{K_{12}}{K_p}$$

In practice, the ageing greatly varies from one vehicle to another, and several parameters seem to influence it, from the type of products transported, to the architecture of the refrigerated units. The purpose of the study was therefore to use the data available for refrigerated vehicles, in order to build a classifier able to predict the ageing of a vehicle based on its characteristics.

In addition, the ageing is a continuous attribute, but once computed, it was divided into two categories: high or low ageing. Indeed, the exact value itself is not necessarily important for domain expert, and being able to predict a categorical variable turns a regression problem into a regression one, while giving a very useful indication to the domain experts.

2.2 Refrigerated Transport Vehicle's Data

At 31 December 2017, based on the Datafrig® data, the French fleet counted 110 000 refrigerated transport equipment with a valid ATP certification. These equipments are divided into different categories, the main ones are: vans (vehicles with a total weight allowed in charge smaller than 3.5 tons), trucks (vehicles whose total weight allowed in charge varies from 3.5 to 29 tons) and semi-trailers (vehicles with total weight allowed in charge higher than 29 tons). The Datafrig® database is managed by Cemafroid and contains all the ATP in-service equipment in France. The Datafrig® data are first self-controlled by the body manufacturer, secondly automatically controlled and analyzed by the Datafrig® software, then checked by a first technician and finally validated by a supervisor. During the audit of the manufacturing of the refrigerated vehicle an independent auditor verifies the quality of these data and compares them to the real equipment. These five steps quality control system are supposed to give to these data high level of quality.

From this database, sample of only 1158 highly curated data could be extracted from the 109 122 refrigerated equipment registered in the Datafrig® database. These 1158 data represent different in-service vehicles of different firms, tested after twelve years in the laboratory of Cemafroid. This is a surprisingly low number, showing that there is only a small available proportion of available data, meaning it has to be as qualitative as possible to hopefully build a useful predictive model.

For each of the vehicles more than 80 features are known. Ten of the most important features, from a domain expert point of view, were analyzed including the K_{12}, the corresponding K_p, the type of vehicle, the body manufacturer, the nature of the refrigerated enclosures and their isolation, the use of the vehicles and the different types of transport to which they may be subjected. Most of the available features are therefore categorical. Our selection criteria for the sample of 1158 vehicles were based on data quality requirements applied on the ten selected attributes (or variables): no null values, no outliers, and no duplicates. It is worth mentioning that we rely on knowledge domain provided by experts to identify our ten features. A statistical analysis of this dataset is available in [7].

2.3 Initial Prediction Results

Once the cleaned dataset obtained, a traditional classification workflow was applied: the data was split in a training (80%) and a testing set (20%). The objective was to sketch the construction of a first initial model, before any analysis of the dataset, using classical state-of-the-art methods. It was decided to build a first classification model using a decision tree [6], in order to obtain an interpretable model, that could easily be discussed with thermal engineers. To improve the results, a boosted version of the algorithm (see [19]) was used. The results were very encouraging, with a precision of 0.818 and a recall of 0.783. The F1 score was of 0.800. In addition, a ROC curve of this first classifier is presented on Fig. 1, also showing good performances for the ageing prediction.

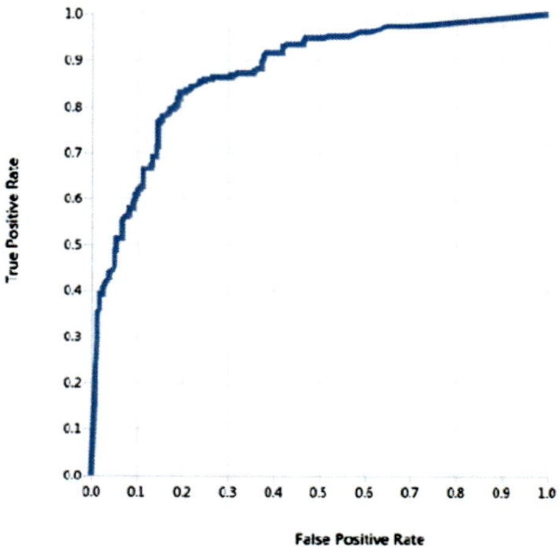

Fig. 1. ROC curve for the prediction of refrigerated vehicles ageing

With this first model available, it was possible to engage in a discussion with the experts in thermal engineering. The purpose was to understand what these result meant, but also and especially what could be done to improve the results, and what were the blocking points in the dataset itself. For example, some tuples might require additional cleaning, or some records in the database might cause problem for the learning process. As checking the data manually is very tiresome, even on a small dataset, it appeared that the notion of functional dependency could be extremely useful to assist the data expert in this specific task.

3 Domain Experts and Trust in the Classification Model

3.1 Preliminaries

We first recall basic notations and definitions that will be used throughout the paper. It is assumed the reader is familiar with databases notations (see [12]).

Let U be a set of attributes. A relation schema R is a name associated with attributes of U, i.e. $R \subseteq U$. A database schema $\mathcal{R}$ is a set of relation schemas.

Let D be a set of constants, $A \in U$ and R a relation schema. The domain of A is denoted by $dom(A) \subseteq D$. A tuple t over R is a function from R to D. A relation r over R is a set of tuples over R. If $X \subseteq U$, and if t is a tuple over U, then we denote the restriction of t to X by $t[X]$. If r is a relation over U, then $r[X] = \{t[X], t \in R\}$. The active domain of A in r, denoted by $ADOM(A, r)$, is the set of values taken by A in r. The active domain of r, denoted by $ADOM(r)$, is the set of values in r.

We now define the syntax and the semantics of a FD.

Definition 1. Let R be a relation schema, $X \subseteq R$ and $C \subseteq R \backslash X$. A FD on R is an expression of the form $R : X \rightarrow C$ (or simply $X \rightarrow C$ when R is clear from context)

Definition 2. Let r be a relation over R and $X \rightarrow C$ a FD on R. $X \rightarrow C$ is satisfied in r, denoted by $r \models X \rightarrow C$, if and only if for all $t_1, t_2 \in r$, if $t_1[X] = t_2[X]$ then $t_1[C] = t_2[C]$.

Without loss of generality, the scope is limited to crisp FDs, i.e. using only the strict equality. However, many other extensions have been proposed (see for example [5]). In addition, this paper only uses canonical FDs, that only have a unique attribute in their right hand side.

Strictly speaking, it is only possible for a relation to satisfy entirely a FD: otherwise, the dependency is not satisfied, there is no middle ground. When dealing with real life data, it is very likely that the FD will not be satisfied, even though a classification model might then be built with satisfying performances. However, understanding what tuples prevent the FD from being satisfied can then lead to improve the results, if the identified blockages can be explained and fixed by the domain experts. Such tuples are called counterexamples, and are defined as follows:

Definition 3. Let r be a relation over R and $X \rightarrow Y$ a FD f on R. The set of counterexamples of f over r is denoted by $CE(X \rightarrow Y)$ and defined as follows:

$$CE(X \rightarrow Y, r) = \{(t_1, t_2) | t_1, t_2 \in r, t_1[X] = t_2[X] \text{ and } t_1[Y] \neq t_2[Y]\}$$

Counterexamples identify pairs of tuples for which the classifier will never be able to perform correctly, as for the same input, it will always predict the same output. These pairs are therefore very important, as their number directly impacts the quality of the classification. As a result, in order to evaluate the impact of counterexamples on classification, it is necessary to know their proportion in the dataset. Indeed, if a classifier only contains a few counterexamples, the impact on the classification will be marginal. On the opposite, a large counterexample set will significantly impact the accuracy results.

Evaluating the impact of counterexamples can be a little subtle. Indeed, a single tuple might cause many counterexamples, if it is in conflict with many other tuples that agree between them. On the opposite, on other relations, the counterexamples might be all due to many different tuples that each are in conflict with only a few other tuples. This problem is actually equivalent to estimating the error of the FD in a relation, a problem addressed in [10], in which three measures are presented, given a FD $X \rightarrow C$ and a relation r.

The first one, G_1, gives the proportion of counterexamples in the relation:

$$G_1(X \rightarrow C, r) = \frac{|\{(u, v) | u, v \in r, u[X] = v[X], u[C] \neq v[C]\}|}{|r|^2}$$

Following this first measure, it is also possible to determine the proportion of tuples involved in counterexamples. This measure G_2 is given as follows:

$$G_2(X \to C, r) = \frac{|\{u | u \in r, \exists v \in r : u[X] = v[X], u[C] \neq v[C]\}|}{|r|}$$

These two metrics are designed to evaluate the importance of counterexamples in the relation. Similarly, measure G_3 computes the size of the set of tuples in r to obtain a maximal new relation s satisfying $X \to C$. Contrary to [10] that present this measure as an error, we propose it as follows:

$$G_3(X \to C, r) = \frac{max(\{|s| | s \subseteq r, s \models X \to C\})}{|r|}$$

In [11], it is underlined that G_3 is a direct upper bound for the accuracy of a classifier trained on the considered data, when the FD between the features and the class is considered, exactly as for the application to CEMAFROID's data.

3.2 Presentation of Counterexamples

Table 1. Subset of counterexamples from the classification dataset, on the ageing of refrigerated transport vehicles.

id	Manufacturer	Cell type	Insulation type	Vehicle type	Products	Ageing t_1	Ageing t_2
1	Firm 1	Integrated	Polyuréthane	Truck	Meat	High	Low
2	Firm 2	Integrated	Polyuréthane with cylopentane	Van	Fruits	High	Low
3	Firm 3	Rapportee	Polyuréthane with cylopentane	Panel truck	Frozen food	High	Low
4	Firm 4	Integrated	Polyuréthane with cylopentane	Trailer	Vegetables	High	Low
5	Firm 5	Rapportee	Polyuréthane with cylopentane without CFC	Truck	Cheese	High	Low
6	Firm 6	Rapportee	Polyuréthane with cylopentane	Remorque	Dairy products	High	Low

Using the tools presented before, the classification dataset was analyzed to discuss with the experts. A toy sample of counterexamples is presented in Table 1. For sake of clarity, only five attributes are represented in this example. Each line represents a counterexample, which is two tuples. As they both share the same values over the classification features, only the ageing column, on which they differ, is represented for both (Ageing t_1 and Ageing t_2). This is also how the counterexamples were presented to the thermal engineers, using an interface developed to present the counterexample. It is named LeaFF (Learning Feasibility with FDs), and can be used to query the data used for prediction, retrieve the counterexample, and obtain the metrics for FD satisfaction. A snapshot of this interface is presented on Fig. 2, applied to CEMAFROID's data: it shows the part of the interface that can be used to enter the features and the class, so that the FD can be checked and counterexamples retrieved if necessary.

Based on such a table, they examined the counterexamples, to see if they could explain their existence. We created a web interface to retrieve and present the counterexamples. The part of this interface dedicated to the counterexamples is presented on Fig. 2: a zone allows to specify the left and right part of the dependency. It is then checked, and if not valid, counterexamples and metrics are presented below.

3.3 Measuring the Counterexamples Rate

The three aforementioned metrics were computed over the refrigerated vehicles dataset. The proportion of counterexamples, $G_1 = 9.02\%$, is low, showing that the pairs of tuples in the dataset are not a big proportion. However, as $G_2 = 100\%$, all tuples from the dataset are involved in at least one counterexamples: this is likely because a few tuples are in conflicts with almost all the other. This is confirmed by measuring $G_3 = 86.73\%$: this shows that to obtain a counterexamples-free dataset, the vast majority of the data can be saved.

When presented with the measures, the domain expert had many questions regarding what exactly each measure meant, and how to know whether the results were good or not: is it good if the score is low? Or should it be higher to be better? Indeed, these measures are not trivial, and the meaning they carry over the pairs of tuples is not intuitive right the first time. This interaction therefore highlighted that it will be interesting to find a convenient visualization for domain experts, to present this result in a more instinctive manner.

3.4 Explaining Counterexamples

After a thorough discussion, it appeared that several factors could explain the counterexamples, each having a different solution. Ultimately, these factors were divided into three main categories.

Dirty Data
A considerable amount of time had been spent on preparing and cleaning the dataset in order to use it for classification. However, when looking at counterexamples, it appeared that some tuples contained data that appeared to be incorrect. For counterexample 1 in Table 1, it appeared that the vehicle with a low ageing was actually a van instead of a truck. Similarly, for counterexample 2, the vehicle with a high ageing actually transported meat instead of vegetables. For this first category, once the data experts were aware of their existence, it was easy for them to spot these mistakes, based on the value of all the attributes. In addition, these counterexamples are easy to fix by correcting the wrong values.

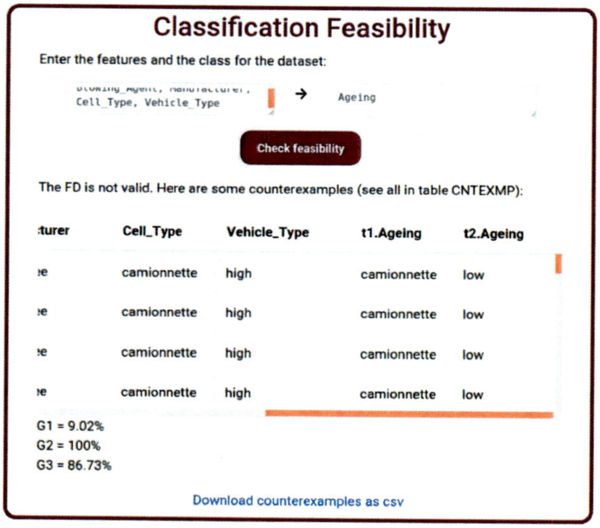

Fig. 2. Snapshot of the interface LeaFF used to explore FDs and counterexamples

Missing Information

For other counterexamples, it appeared that the attributes selected for classification were not enough to explain their existence. However, other attributes, that had not been kept for classification, allowed to discriminate between the two tuples involved in the counterexample. For instance, in Table 1, counterexample 3 can be removed if the number of food cases in the vehicle is taken into account. For counterexample 4, a specific characteristic of the cooling unit differed between the two tuples. As a result, taking those additional attributes, that at first had not been considered relevant for the classification, allowed to remove counterexamples, that will then improve the classifier's performances. This is a way to do feature engineering in collaboration with the domain experts.

Human Factor

Finally, for a last group of counterexamples such as number 5 and number 6 from Table 1, it appeared that the only explanation was a human factor, such as how the driver operates the vehicle. Indeed, this is susceptible of influencing the ageing of the vehicle, but it is also hard to quantify. As a result, this last class of counterexamples is very difficult to fix, as data cannot be cleaned or completed. However, being aware that such counterexamples exist helps the data expert in understanding the limitations of the classification model being produced. Finally, it also indicates other values that could be interesting to record: in this case for example the average speed of the vehicle. Similarly, during its life, the vehicle is subject to accidents of which nothing is known. Information about the nature and severity of such accidents could be useful for the study of ageing, and could also explain some counterexamples.

Presenting the counterexamples to domain experts therefore allowed to start a very interesting discussion. Not only did it allow to better explain the limitations of the classifier, but it also allowed to find concrete solution to improve the initial dataset in order to improve the classification performances.

3.5 Take Away Lessons

Using the functional dependency to explore the classification dataset has proven to be a very enriching experience, that created fruitful discussions and results. First, the counterexamples allowed to identify limitations in the dataset, and to take concrete actions to get higher quality data, and therefore improve the results for a future new model. The counterexamples are a powerful notion to avoid the domain expert from being overwhelmed by the data, as she then only have a small but meaningful subset of tuples to study. The counterexamples are therefore a perfect starting point for a discussion between data scientist and domain expert: while the first gain knowledge on data they are not expert on, the others can point out important information more easily. The counterexamples are a way for domain expert to read a concrete information that can have an impact on her day-to-day work. In this study, the analysis of the counterexamples also allowed to identify three different concrete problems that can appear in a classification dataset, as well as solutions to address some of them. This study also shows that in classification, it is not possible to ignore the field reality, and to only see the problem as a matrix of data: the physical model is as important as the data itself.

Finally, functional dependencies could also identify situations in which it does not make any sense to look for a classification model: it was not the case for this specific problem, but in other situations, if the ratio of counterexamples is too high, it is hopeless to seek for a good precision score when a significant amount of the data does not seem to follow a mathematical function. The scores presented to the expert are therefore also a way to give assurances on the model, to be aware of both its limitations and strengths.

4 Related Work

The analysis of counterexamples, and the eventual removal of some of them, for example through contextualization, is a form of data cleaning, as it aims at removing the data that is not fitted for the considered task. Data cleaning is a crucial part in most of data science application, as data scientist actually spend around 80% of their time on cleaning the data [21]. As a consequence, many research has be done on addressing this problem [15], many of them relying on data dependencies and constraints. For instance, [20] proposes a semantic data profiler that can compute samples that satisfy the same constraints than a given dataset. As the limited expressiveness of functional dependencies did not always adapt well to the need of data cleaning on real datasets, specific dependencies have been proposed to identify inconsistencies in a dataset, and eventually repair it. Conditional dependencies [3] are functional dependencies that hold only on a subset of the dataset. Matching dependencies [9] for data repairing uses matching rules to relax the equality on functional dependencies and assign values for data repairing. In Holoclean [16], dependencies are used to clean automatically a dataset. In [17], a formal framework is proposed to bridge the gap between database theory and learnability theory, and is applied to three applications: data cleaning, probabilistic query answering, and learning. It can even be used to clean dataset in order to provide *fairness* [18]. [8] introduces denial constraints, allowing to declaratively specify logical formulae to exclude counterexamples.

5 Conclusion

This paper presents the results obtained to better trust classification models in the context of a study with CEMAFROID, a french company delivering conformity attestations for refrigerated transport vehicles. To this end, a classification dataset was built and a first classifier was elaborated using standard classification methods. It showed very promising results, but also raised interrogations from the thermal engineers, related to trust issues in machine learning. In order to understand the dataset and to demystify the classifier, we proposed to see the classification model from the features to the class to be predicted as a functional dependency. The counterexamples to this FD then give direct limitations of the classifier. The counterexamples were analyzed, and various causes for their existence were discovered, as well as potential solutions to remove them. More importantly, domain experts were able to better understand the pros and cons of the numerical model, influencing their trust in the whole machine learning process. This study was therefore really enriching, allowing to bridge a gap by using FDs, a well-known notion in databases, into a machine learning process. The methodology from this paper could be used in any classification problem, to first study the existence of a model before trying to learn from data.

This study also raises many open research questions, challenging but highly relevant for any classification problem. First, if we only considered crisp functional dependencies in this paper, it could be interesting to refine the comparability between tuples, in order to mange each attribute at a finer level, for example with similarity measures. Then, if the feedback obtained during this study is highly compelling, the next step would be to use it to clean the entire dataset, before building a new classification model, hopefully reaching higher performances. This requires another step to automate the cleaning process of the dataset, for example using the chase algorithm [1].

References

1. Aho, A.V., Beeri, C., Ullman, J.D.: The theory of joins in relational databases. ACM Trans. Database Syst. (TODS) **4**(3), 297–314 (1979)

2. Armstrong, W.: Dependency structures of database relationship. In: Information Processing, pp. 580–583 (1974)
3. Bohannon, P., Fan, W., Geerts, F., Jia, X., Kementsietsidis, A.: Conditional functional dependencies for data cleaning. In: IEEE 23rd International Conference on Data Engineering 2007, ICDE 2007, pp. 746–755. IEEE (2007)
4. Boldrin, B., Minotto, G., Panozzo, G., Toniolo, B., Varotto, G.: Ageing of insulated vehicles. statistical elaboration of tests results. Boldrin, B. et al., 1990–9 (1990)
5. Bosc, P., Dubois, D., Prade, H.: Fuzzy functional dependencies and redundancy elimination. J. Am. Soc. Inf. Sci. **49**(3), 217–235 (1998)
6. Breiman, L.: Classification and Regression Trees. Routledge, Abingdon (2017)
7. Capo, C., Petit, J.M., Revellin, R., Bonjour, G., Cavalier, G.: Ageing of in-service refrigerated transport vehicles: a statistical analysis. In: 25th IIR International Conference on Refrigeration (2019)
8. Chu, X., Ilyas, I.F., Papotti, P.: Discovering denial constraints. Proc. VLDB Endowment **6**(13), 1498–1509 (2013)
9. Fan, W.: Dependencies revisited for improving data quality. In: Proceedings of the Twenty-Seventh ACM SIGMOD-SIGACT-SIGART Symposium on Principles of Database Systems, pp. 159–170. ACM (2008)
10. Kivinen, J., Mannila, H.: Approximate inference of functional dependencies from relations. Theor. Comput. Sci. **149**(1), 129–149 (1995). https://doi.org/10.1016/0304-3975(95)00028-U
11. Le Guilly, M., Petit, J.M., Scuturici, V.M.: Evaluating classification feasability over datasets using functional dependencies. In: BDA 2019 35ème conférence sur la Gestion de Données: Principes, Technologies et Applications, Lyon, France (2019)
12. Levene, M., Loizou, G.: A Guided Tour of Relational Databases and Beyond. Springer, Heidelberg (1999)
13. Panozzo, G., Alberti, O., Toniolo, B., Barizza, A., Boldrin, B.: Parameters affecting the ageing of insulated vehicles. In: Proceedings of the 20th International Congress of Refrigeration IV (1999)
14. Panozzo, G.B.B., Minotto, G., Toniolo, B., Biancardi, N.: Ageing of insulated vehicles: theoretical model and experimental analysis. In: Proceedings of the 19th International Congress Refrigeration, Den Hague, Netherlands, vol. II, pp. 583–589 (1995)
15. Rahm, E., Do, H.H.: Data cleaning: problems and current approaches. IEEE Data Eng. Bull. **23**(4), 3–13 (2000)
16. Rekatsinas, T., Chu, X., Ilyas, I.F., Ré, C.: Holoclean: holistic data repairs with probabilistic inference. Proc. VLDB Endowment **10**(11), 1190–1201 (2017)
17. Sa, C.D., Ilyas, I.F., Kimelfeld, B., Ré, C., Rekatsinas, T.: A formal framework for probabilistic unclean databases. In: 22nd International Conference on Database Theory, ICDT 2019, Lisbon, Portugal, 26–28 March 2019, pp. 6:1–6:18 (2019). https://doi.org/10.4230/LIPIcs.ICDT.2019.6
18. Salimi, B., Rodriguez, L., Howe, B., Suciu, D.: Interventional fairness: causal database repair for algorithmic fairness. In: Proceedings of the 2019 International Conference on Management of Data, pp. 793–810. ACM (2019)
19. Schapire, R.: The strength of weak learnability. Mach. Learn. **5**(2), 197–227 (1990)
20. Wei, Z., Link, S.: DataProf: semantic profiling for iterative data cleansing and business rule acquisition. In: Proceedings of the 2018 International Conference on Management of Data, pp. 1793–1796. ACM (2018)
21. Zhang, S., Zhang, C., Yang, Q.: Data preparation for data mining. Appl. Artif. Intell. **17**(5–6), 375–381 (2003)

Perspectives on Artificial Learning

Automatic Stopwords Identification from Very Small Corpora

Stefano Ferilli(✉), Giovanni Luca Izzi, and Tiziano Franza

University of Bari, Bari, Italy
stefano.ferilli@uniba.it

Abstract. Natural Language Processing tools use language-specific linguistic resources, that might be unavailable for many languages. Since manually building them is complex, it would be desirable to learn these resources automatically from sample texts. In this paper we focus on *stopwords*, i.e., terms which are not relevant to understand the topic and content of a document. Specifically, we compare the performance of different techniques proposed in the literature when applied to very small corpora (even single documents), as may be the case for very local languages lacking a wide literature. Experiments show that simple term-frequency is an extremely reliable indicator, that outperforms other more complex approaches. While the study is conducted on Italian, the approach is generic and applicable to other languages.

Keywords: Information Retrieval · Natural Language Processing · Linguistic resources · Stopword extraction

1 Introduction and Motivation

Natural Language Processing (NLP) aims at developing tools for understanding the components, structure and meaning of texts, and to properly organize this information so as to help human users in satisfying their information needs. Most NLP tasks are carried out using language-dependent linguistic resources (thesauri, lists of words or suffixes, processing tools, etc.). Since languages are very different from each other, the resources cannot be reused across languages. While for some languages, such as English, many high-quality resources have been developed, little exists for a few other widespread languages, and almost nothing for the vast majority of languages having limited diffusion or not used for mainstream communication all over the world. As a result, NLP cannot be applied to documents in these languages. Since manually developing linguistic resources for a language is a complex activity to carry out manually, a possible solution is to learn the resources automatically from texts in that language.

Some NLP tasks are often used for pre-processing of texts in the context of other tasks and applications. One application that requires a pre-processing of texts written in natural language is Information Retrieval (IR for short), where the aim is indexing a corpus of texts so as to quickly and easily find

M. Stettinger et al. (Eds.): ISMIS 2020, SCI 949, pp. 31–46, 2021.
https://doi.org/10.1007/978-3-030-67148-8_3

and return the most relevant ones to a query expressed as a set of terms. In particular, an important pre-processing step for IR approaches at the lexical level consists in identifying and removing *stopwords*, defined in [7] as "terms that occur most frequently in a document and contain very little information that is usually not necessary". Discussing indexing of technical literature by most significant words ('keywords'), [9] defines stopwords by contrast as follows: "Since significance is difficult to predict, it is more practical to isolate it by rejecting all obviously non-significant or "common" words, with the risk of admitting certain words of questionable status. Such words may subsequently be eliminated or tolerated as so much "noise". A list of nonsignificant words would include articles, conjunctions, prepositions, auxiliary verbs, certain adjectives and words such as "report," "analysis," "theory," and the like".

In IR, the stopword removal task is aimed at reducing the dimensionality of the problem (this way increasing speed and reducing storage requirements) and at improving the quality of the statistics on which indexing is based (this way improving accuracy). It is typically carried out by look-up in a pre-determined list of words, known as 'stopword list'. Since stopwords only have a syntactic function, one way for preparing stopword lists is including all 'function words', i.e. terms associated to invariant Parts-of-Speech of the language (e.g., articles, conjunctions, pronouns and prepositions—see [9]). This approach has several drawbacks:

1. it requires prior knowledge about the grammar of the language;
2. stopwords are domain-dependent, and for domain-specific applications also other terms, that are insignificant in the particular context, might be added to the list (e.g., in the example by [9], 'report', 'analysis' and 'theory' are considered as irrelevant in the domain of technical literature);
3. manually defining suitable stopword lists is costly (it requires linguistic experts), time-consuming (much study and analysis is required), and error-prone (as all human activities).

So, in this paper we focus on the automatic learning of stopword lists, with the specific aim of investigating the effectiveness of different techniques under the following constraints:

- using just plain texts in a given language for learning;
- not using any grammatical knowledge;
- working on *very small* corpora (less than a dozen texts);
- being fully automatic;
- being generally applicable (not tailored to a specific language)[1].

These constraints are typically encountered when working on local languages, for which very few texts and no linguistic studies are available, such as dialects,

[1] Of course, we mean applicable to languages having the same lexical and syntactic structure as English, Italian, etc. E.g., it would not be applicable to Vietnamese, where words are written with a space between each syllable. On the other hand, it might be applicable to inflected languages, assuming that inflected stopwords are sufficiently frequent so as to be selected by the algorithm.

jargons or languages spoken by linguistic minorities and in danger of extinction (with the consequence of a huge cultural loss). Being able to automatically index the few available texts in these languages would support their accessibility and save them from oblivion. Also, further texts that might become available in the future might be immediately indexed using the stopword lists learned on the few initial texts.

In particular, we consider the extreme case of text corpora consisting of just one document, under different linguistic conditions. As far as we know, this is an original setting, since all other studies in the literature consider huge amounts of data, which are not always available in practice. Following are some practical cases that we actually faced in our work, and in which our approach was needed, which inspired the investigation reported in this paper:

1. In Italy, almost every single town has its own dialect, which is an actual language in all respects, and may be significantly different even from those of neighboring towns. These languages mostly live in oral conversation, and a few written texts are available from the past. However, due to publishing being much more easily carried out in recent years, thanks to modern technologies, and thanks to the increasing interest in rediscovering and saving such languages, many new texts in these dialects are being published. Automatically indexing such texts, so as to make them readily available for term-based querying, requires stopword removal. However, linguistic studies that provide stopword lists manually built by experts are not available, and automatic learning of such lists must be based on the very few texts available so far.
2. Any specific field of interest, or specific mix of topics of interest, requires its own stopword list, but only generic stopword lists are readily available for use. In order to index document collections specifically built about such topic(s), tailored stopword lists must be available for carrying out stopword removal. Again, the manual approach to building such stopword lists is infeasible in most cases. On the other hand, it would be useful to learn such stopword lists on a few representative documents, and then use them to index the whole collection.

In fact, we will show that, under the above constraints, the simplest approach, based on plain frequency count of terms, is much more effective than significantly more complex approaches. This is a non-trivial finding, which raises the question about when more complex approaches should be used to obtain real advantages in terms of effectiveness. We ran our experiments on Italian, as a sufficiently complex language for making us confident that similar experimental results may be obtained for other languages, as well.

After discussing related works in the next section, Sect. 3 reports our experimental setting and describes the dataset used, while Sect. 4 proposes our experimental study of stopword extraction in single texts or small corpora. Finally, Sect. 5 concludes the paper and outlines future work directions.

2 Related Work

A quick overview of stopword removal algorithms can be found in [7]. Due to the relevance of stopword removal for IR, many works were not explicitly evaluated from the linguistic perspective, but indirectly evaluated the quality of the extracted stopwords from the retrieval performance.

A number of works available in the literature, and focused on specific languages (e.g., Arabic [1], Chinese [17], Punjabi [10] or Hindi [6]) will not be commented in detail here, since we aim at a general approach that can be applied to different languages.

Working in an IR setting, [15] uses a Vector Space Model to identify stopwords. However, the proposed technique applies stemming prior to the stopword extraction step, which makes the approach language-dependent, and requires, again, the existence of tools/resources for that language (which we want to avoid).

Two more purely frequency-based approaches are proposed in [5,12]. The former still deals with English, and was tested on a corpus of broad literature including more than 1 million words. The latter specifically focuses on French, and was applied to two corpora made up of many small texts, but totaling more than 4 and more than 6 million words, respectively. Moreover, both manually adjust the automatically determined list of stopwords.

[13] introduced a method for optimizing a stopword list by combining stochastic search (hill-climbing and evolutionary algorithm) with k-means clustering. The lists are obtained based on the entropy of words calculated on the Bank Search dataset and a random data dataset. A series of experiments obtained a median performance value of 93.05% of Van Rijsbergen's stopword list.

Lo et al. [8] propose the Term-based Random Sampling (TRS) approach, based on the Kullback-Leibler (KL) divergence measure to determine if a word contains useful information, and compare its performance in supporting IR tasks to those of four variations of an approach based on Zipf's law[2]. While the former is more complicated than the latter, both proved to be very effective. The four variations differ from each other for the measure used to determine the degree of 'stopwordness' for candidate stopwords: Term Frequency (TF), Normalized Term Frequency (NTF), Inverse Document Frequency (IDF) and the Normalized Inverse Document Frequency (NIDF). They obtained the best performances using NIDF. The TF approach was also implemented in **BLA-BLA** [4], a tool to learn in a fully automatic way linguistic resources, and used in [3] to show that standard lists of stopwords widely used in the literature carry significant biases that may tamper their effectiveness. For this reason, and due to their being language-independent, we will use these techniques in our comparison experiments, and will describe them in more details in the following.

Let us first define some notation. Let us call $\mathcal{C}$ the available corpus of documents in a given language, to be used for learning a stopword list for that

[2] By Zipf's law, the distribution of frequency of terms rank can be described very precisely by the relation $F(r) = \frac{C}{r^\alpha}$ where $\alpha \approx 1$ and $C \approx 0.1$.

Algorithm 1. Baseline algorithm [8]

Require: vocabulary V for $\mathcal{C}$; threshold θ
  for all $t \in V$ **do**
    if $f(t) \geq \theta$ **then**
      remove t as a stopword
    end if
  end for

language. Call $n = |\mathcal{C}|$ the number of documents in $\mathcal{C}$, and $V = \{t_1, \ldots, t_m\}$ the corresponding *vocabulary*, i.e., the set of distinct terms appearing in $\mathcal{C}$. Given a term $t_i \in V$, let us call o_i the number of times it occurs in $\mathcal{C}$, and n_i the number of documents in which it occurs. Let us also consider $o = \sum_i o_i$, the overall number of *tokens* (i.e., occurrences of terms) in $\mathcal{C}$. In the following, $\mathcal{C}$ will be considered as fixed, and thus ignored in the notation.

Algorithm 1 shows the 'baseline' algorithm for the approaches based on Zipf's law. The four approaches replace function $f(t)$ by four different measures that will yield, in almost all cases, different lists of stopwords:

- Term frequency (TF): the number of times a certain term appears in a given documents collection:
$$f(t_i) = tf_i = o_i$$
- Normalized Term Frequency (NTF): obtained by normalizing the term frequency with respect to the total number of tokens in the collection:
$$f(t_i) = ntf_i = -\log(\frac{o_i}{o})$$
- Inverse Document Frequency (IDF) [14]: using the frequency distribution of the terms with respect to the number of documents:
$$f(t_i) = idf_i = \log(\frac{n}{n_i})$$
In other words, low-frequency terms are more likely to appear in relevant documents and should be considered more informative and therefore more important within these documents.
- Normalized IDF (NIDF): the most common form of IDF weighing normalizes the IDF with respect to the number of documents that don't contain the term $(n - n_i)$, adding a constant 0.5 to both the numerator and the denominator to mitigate extreme values [11]:
$$f(t_i) = nidf_i = \log(\frac{(n - n_i) + 0.5}{n_i + 0.5})$$

TRS is described in Algorithm 2. Its basic idea is that the more important/informative a term, the less likely it is a stopword, because stopwords do not contain any information. So, this approach should yield a different set of

Algorithm 2. Term-based random sampling [8]

Require: vocabulary V for $\mathcal{C}$
Require: n: number of attempts
Require: m: number of terms (candidate stopwords) to select at each attempt
Require: l: number of stopwords to be returned
  for $i = 1, \ldots, n$ **do**
    Randomly choose a term t_i in V
    Retrieve the set D_i the documents in $\mathcal{C}$ that contain t_i
    Assign a weight to every term in D_i using the Kullback-Leibler divergency measure
    Divide each term's weight by the maximum weight among all terms
    Add the m least weighted terms to the list of candidate stopwords
  end for
  Shrink the list by merging the elements containing the same term and take the average of the term's associated weights
  return l least weighted terms as stopwords

stopwords than the Baseline variations. It randomly selects n terms (to ensure a better coverage of the entire collection). Each such term, that might be a stopword or a really informative term, determines a set of candidate stopwords as follows[3]: all the documents containing the term are collected in a sample, and the importance of each term in the sample[4] is assessed by comparing its distribution within the sample and within the whole collection of documents, using the KL divergence measure [2]:

$$d_x(t) = P_x(t) \cdot \log_2 \frac{P_x(t)}{P(t)}$$

where:

- $P_x(t)$ is the normalized frequency of term t within a set of sample documents;
- $P(t)$ is the normalized frequency of term t within the whole collection.

Finally, after shrinking of the list of candidate stopwords returned by each term, the last informative candidates are returned as stopwords. The experiments suggested that the best values to be assigned to the parameters m, n and l are 200, 1000 and 400 respectively.

[3] According to [8], this approach is very similar to the one used in [16] for query expansion in IR by finding terms that have the same or similar meaning as a given term.

[4] Note that, if the term rarely occurs in the collection, the retrieved set of terms would be small. E.g., a term occurring in just one document would return only the other terms in that documents. Selecting n terms should overcome the problem and yield a better sample that allows a better estimation of the distribution and importance of terms.

3 Experimental Setting

In our experiments, we consider documents represented as plain text, each associated to the corresponding language. We assume that each document belongs exactly to one language: of course, it may include words or expressions from other languages, but these are to be considered as noise, and suitably handled by the learning approaches. We define a *word* as a sequence of alphabetic characters only, delimited by non-alphabetic characters.

We want to study the behavior of different approaches when only a *very* small corpus is available for learning (say, made up of a dozen texts). This should obviously stress the learning approach, because we may expect that, processing a large number of texts, the frequency of real stopwords will in the end become clearly predominant over the other words.

As said, we focused on Italian, as a language that has attracted some attention from the NLP community (albeit not as much as English), so that established stopword lists exist that may serve as a golden standard for our experiments. Italian has a more complex structure than English, not only from a syntactic viewpoint, but also from a morphological one (inflection is much richer than in English, and this might have an impact on frequency-based stopword identification). So, we expect that, if good results are obtained on Italian, then good results might be obtained on many other languages, as well.

We purposely created a dataset for our study, by selecting texts that cover a wide range of styles (fiction, technical, poetry), genres (novels, tales, history, law, allegorical, medicine) and ages (ranging from the XIV to the XXI century), so as to analyze different perspectives on the language. This allowed us to check the effectiveness of stopword removal techniques in different cases:

- single texts in different styles;
- single texts in different lengths;
- aggregations of texts in the same style;
- aggregations of texts in the different styles.

Indeed, we posit that the use of texts in different styles should improve the quality of the results, because real stopwords should emerge more clearly, while domain- or topic-specific stopwords should have less impact overall. This is also a novel setting, since the literature focused so far on homogeneous collections, and many are skeptical about the use of mixes of styles, especially in the case of so few documents for each style.

Specifically, we selected the following texts[5]:

La Divina Commedia the allegorical/didactic poem by Dante Alighieri;

[5] The texts are the same as in [3], with the addition of HeG and AdA. So, performance reported in Table 4 for the TF approach on the other single texts is the same as in [3], as well. However, due to the addition of HeG and AdA, performance reported in Table 4 for the TF approach on NTT and All has changed with respect to [3]. On the other hand, all performances reported for the other approaches, and their comparison, are presented in this paper for the first time.

Table 1. Statistics on the processed texts

ID	Text	Chars	Words	Vocabulary	Year
LDC	La Divina Commedia	561149	97714	12796	1321
CCI	Codice Civile Italiano	1511666	228251	8659	1942
L'E	L'Esclusa	337589	55846	8919	1908
IPS	I Promessi Sposi	1307423	220174	19658	1842
TlN	Tutte le Novelle	1591823	264703	21641	1877–1922
PPI1	Passeggiate per l'Italia 1	438868	71467	11995	1856–1877
PPI2	Passeggiate per l'Italia 2	549884	86818	14710	1856–1877
PPI3	Passeggiate per l'Italia 3	478110	75871	12721	1856–1877
PPI4	Passeggiate per l'Italia 4	472272	75618	12183	1856–1877
PPI5	Passeggiate per l'Italia 5	289006	46655	10470	1856–1877
HeG	Hansel e Gretel	14970	2485	890	1812
AdA	Appunti di Anatomia 1	261237	39287	5456	2009

Codice Civile the Italian Civil Code, a collection of regulations;
L'Esclusa a classic novel by Luigi Pirandello, a novel;
I Promessi Sposi a historical novel by Alessandro Manzoni;
Tutte le novelle a collection of short stories by Giovanni Verga;
Passeggiate per l'Italia by Ferdinando Gregorovius, an account of travels around Italy in 5 volumes;
Hansel e Gretel a fairy tale by the Grimm brothers[6];
Appunti di Anatomia 1 lecture notes on anatomy by Giordano Perin.

Table 1 reports some statistics about the selected texts: length (in number of characters and of words), vocabulary (i.e., the number of different words), and year(s) of publication. It also assigns an acronym (*ID*) to be used as an identifier of the text in the rest of the paper. The number of characters and words is approximate (counted by a text editor), while the size of the vocabulary is exact (computed by the pre-processing step). Most documents have vocabularies of about 10 thousand words, with the notable exception of HeG having a vocabulary of less than 900 words. Due to the presence of texts in different styles and of different length, we believe that this collection of texts can be considered representative enough to draw general conclusions on the behavior of the selected approaches on single texts, or on corpora made up of a handful of texts, from the results of our experiments. Also, since the selected texts are obtained by applying

[6] It is a translation, not an original Italian text. Some might object that translations should not be considered in experiments concerning a language. We do not agree: we believe that translations produced by mother-tongue writers are in any case a direct expression of the target language, and thus can be in all respects considered as target language texts. Moreover, using also translated texts in the experiments may test the effectiveness and robustness of the methodology.

Optical Character Recognition to scanned images of paper books' pages, they contain typos. This allows us to run our experiments on noisy data, which is what one may expect to have in real-world settings.

In order to check the behavior of the approaches under different conditions, we ran our experiments both on single texts and on the following aggregates (including texts in the same style or mixing texts from different styles):

PPI the whole 'Passeggiate per l'Italia', a case of taking several texts in the same narrative style;
NTT the subset of non-technical texts (i.e., PPI, L'E, IPS, HeG, and TlN), a case of taking texts from different kinds of narrative styles;
All the whole set of texts, a case of mixing texts in standard style and several specialized styles.

For performance evaluation, we needed a collection of stopwords to be used as the golden standard. We chose the stopword list provided by Snowball[7], a well-known tool exploited by many NLP systems currently in use. In its complete form, it consists of 279 stopwords.

Performance was evaluated in terms of precision (P) and recall (R). While precision is important to avoid removing false stopwords when pre-processing the text, it is also relevant to check recall, i.e., how many stopwords from the golden standard are retrieved, to understand whether the automatic techniques could be used to automatically provide stopword lists that are similar to what human experts would produce. However, due to the techniques being applied to very small corpora, which might miss some stopwords in the golden standard, we focused more on precision, using recall just as a secondary indicator. To study the overall trend of precision, we measured it when cutting the ranked list of candidate stopwords at various positions n. So, $P@n$ means precision of the top n items in the ranking (i.e., the percentage of the first n items in the list that are also in the golden standard). Specifically, we used values of n that are multiples of 10. We stopped at $n = 100$ both for the sake of readability, and because they represent a 'safety region' where most relevant stopwords should be included, as noted by [3]. As to recall, since in this case we are not interested in the trend, we computed R@100 only. Note that, while R is necessarily monotonic for increasing n, P is not, so some fluctuations are possible.

4 Experimental Study

In this section we report and comment a comparison of three approaches to stopword identification taken from [8], aimed at checking their performance on very small corpora including texts of varied age, length, and style:

- NIDF, selected because [8] reported it to be the best one in its evaluation;
- TRS selected due to its random nature that could identify different (and possibly unexpected) stopwords for different executions, and also because [8] suggests it might be worth more investigation;

[7] http://snowball.tartarus.org/algorithms/italian/stop.

Table 2. Precision of NIDF

Text(s) #	P@10	P@20	P@30	P@40	P@50	P@60	P@70	P@80	P@90	P@100
CCI	0	0	0	0	.02	.06	.08	.07	.06	.07
PPI1	0	0	0	0	0	0	0	0	.03	.05
PPI2	0	0	.03	.02	.04	.07	.10	.09	.08	.07
PPI3	0	.05	.03	.05	.10	.12	.10	.10	.10	.09
PPI4	0	0	0	.02	.02	.02	.01	.01	.03	.05
PPI5	0	0	0	.02	.02	.02	.01	.05	.06	.07
IPS	0	0	0	.02	.02	.02	.04	.06	.08	.07
L'E	0	0	.03	.02	.02	.03	.09	.09	.08	.07
LDC	0	.05	.03	.02	.10	.11	.10	.08	.10	.09
HeG	0	.05	.03	.10	.14	.13	.14	.12	.13	.13
TlN	0	0	.03	.02	.02	.02	.03	.05	.08	.07
AdA	0	0	0	0	0	0	0	.02	.04	.06
PPI	.50	.25	.20	.15	.12	.10	.09	.07	.07	.06
NTT	.50	.25	.20	.32	.42	.43	.43	.42	.38	.34
All	.80	.85	.83	.62	.52	.45	.43	.39	.39	.40

- TF, selected because it follows directly from the definition of 'stopword', and in [3] it obtained good performance on small corpora and was even able to reveal some flaws in existing standard stopword lists for Italian.

Our research hypothesis is that TF may be effective on very small corpora, and actually more effective than more complex approaches, in spite of its extremely simple strategy. According to the experimental setting, we will run the different approaches on increasingly large corpora, starting from single documents up to a dozen document, both in the case of homogeneous texts and in the case of mixed-style texts. Experimental results will be reported in several tables, highlighting in bold the best cases ($P > 0.90$).

Table 2 shows the precision values for NIDF. The stopwords extracted from individual documents are basically useless (precision is almost always basically 0, and never above 0.14). Considering more documents (i.e., document aggregates PPI, NTT and All) precision slightly increases, but anyway never reaching significant (let al.one acceptable) levels for larger values of n. No values are in bold, not even when using all the texts. Given the good experimental results obtained with much more processed text in [8], we must conclude that this approach can be adopted only when a large quantity of documents is available.

As regards the TRS approach, since at each run it randomly chooses the seed word, and thus returns different results, we ran it 5 times on each (set of) document(s), and report the mean precision value in Table 3. Non-monotonic behavior for increasing values of n is evident. Just like NIDF, TRS is totally unsuitable for extracting meaningful stopwords from individual documents.

Table 3. Precision of TRS

Text(s) #	P@10	P@20	P@30	P@40	P@50	P@60	P@70	P@80	P@90	P@100
CCI	0	0	.03	0	.04	.06	.02	0	.02	.02
PPI1	0	0	0	0	0	0	0	.03	.01	.02
PPI2	0	0	.06	.05	.04	.03	.01	0	.01	.03
PPI3	0	0	.06	0	.04	.02	.03	.01	.02	.03
PPI4	0	0	0	.02	.02	.02	.01	.01	.03	.05
PPI5	0	0	0	.02	.02	.02	.01	.05	.06	.07
IPS	0	0	0	.02	.02	.02	.04	.06	.08	.07
L'E	0	0	.03	.02	.02	.03	.09	.09	.08	.07
LDC	0	.05	.03	.02	.10	.11	.10	.08	.10	.09
HeG	0	.05	.03	.10	.14	.13	.14	.12	.13	.13
TIN	0	0	.03	.02	.02	.02	.03	.05	.08	.07
AdA	0	0	0	0	0	0	0	.02	.04	.06
PPI	**.90**	**.90**	**.90**	.85	.84	.82	.74	.67	.62	.58
NTT	**1.00**	**.95**	**.90**	.85	.82	.75	.73	.71	.67	.63
All	.60	**.95**	.53	.70	.76	.41	.44	.40	.38	.33

It performs better on text aggregates (as proven by 7 values in bold and one case of full precision), but with an opposite behavior than NIDF: indeed, it is much better than NIDF when applied to smaller text collections (PPI and NTT), while, surprisingly, on the entire corpus (All) its performance drops, instead of raising and being the best, as one might expect. Overall, we must conclude that its behavior is too variable for drawing general conclusions, and that, again, it is applicable only to large corpora.

Finally, Table 4 reports the precision values obtained by TF. Albeit very simple, and perhaps the most intuitive one, this approach obtains very interesting results, both on single documents, and on document aggregates. Indeed, we may consider its performance as satisfactory on each processed document or document collection at least up to $P@60$. Almost all results for single documents are in bold up to $P@30$, and all are in bold for document aggregates up to $P@50$ (and 2 out of 3 are in bold even in $P@60$). It is much better than the competitors, in spite of their using more complex statistics and/or procedures. The top items in the rank of candidate stopwords are nearly fully correct (precision is 1, or nearly 1, for almost all cases @10, for the vast majority of cases @20, and for the majority of cases @30).

Also noteworthy is the fact that this TF shows a decaying trend in $P@n$ for progressive values of n, while NIDF and TRS had a quite irregular sequence of values, significantly raising or dropping from $P@10$ to $P@100$. The more stable behavior of TF might help when trying to automatically assess the cut point in the ranked list of candidate stopwords, by providing a more reliable

Table 4. Precision of TF

Text(s) #	P@10	P@20	P@30	P@40	P@50	P@60	P@70	P@80	P@90	P@100
CCI	**.90**	**.95**	.87	.88	.76	.68	.61	.58	.54	.53
PPI1	**1.00**	**1.00**	**1.00**	**.95**	**.94**	.88	.83	.79	.73	.73
PPI2	**1.00**	**1.00**	**.97**	**.95**	**.92**	.87	.81	.76	.73	.71
PPI3	**1.00**	**1.00**	**.93**	**.93**	**.90**	**.90**	.83	.75	.73	.68
PPI4	**1.00**	**1.00**	**1.00**	**.93**	.88	.83	.81	.79	.76	.71
PPI5	**1.00**	**1.00**	**1.00**	**.93**	**.92**	.88	.81	.76	.71	.66
IPS	**1.00**	**.95**	**.90**	.85	.80	.73	.73	.70	.67	.65
L'E	**1.00**	**.95**	**.93**	.85	.72	.70	.69	.70	.66	.62
LDC	**1.00**	.85	.83	.80	.74	.67	.64	.60	.58	.53
HeG	**.90**	.70	.70	.70	.68	.65	.61	.56	.53	.52
TlN	**1.00**	**1.00**	**1.00**	**.90**	**.90**	.85	.77	.69	.66	.62
AdA	**1.00**	.85	.73	.62	.58	.55	.50	.45	.43	.41
PPI	**1.00**	**1.00**	**.97**	**.95**	**.94**	**.90**	.86	.83	.78	.72
NTT	**1.00**	**1.00**	**1.00**	**.97**	**.92**	**.90**	.86	.79	.74	.69
All	**1.00**	**1.00**	**1.00**	**.95**	**.92**	.88	.84	.80	.74	.72

basis for techniques based on the progression of values, and allowing a better understanding of the 'stopwordness' ranking.

So, let us analyze in more detail the decaying trend in $P@n$ performance for increasing values of n on specific texts. AdA, LDC and CCI show a faster decay. For AdA it may be explained by the fact that it is not a unitary text, just a collection of lecture notes. So, it is mostly made up of schematic sentences using only the essential words rather than articulated speeches, and thus it includes only a few strictly necessary stopwords. LDC is a poem and it is written in archaic Italian dating back to the 1300 s, so many frequent terms are actually stopwords, but truncated for poetry and missing in the golden standard. Note that, in Italian, even in everyday language some stopwords are truncated: so, this is a further confirmation of the incompleteness of the golden standard noted in [3], rather than an issue with the specific text or style. Finally, CCI contains mostly technical verbs in infinitive form, which are not general stopwords (but, as shown in [3], can be considered as domain-specific stopwords). Some additional comment is worth making about AdA and HeG. $P@n$ performance of the former is the best for lower values of n, but quickly decreases, up to being the worst @100. As regards the latter, it is the worst for lower values of n, but thanks to the smoother decay, @100 it ends up at values close to LDC and CCI, in spite of its being an extremely short text.

Comparing Tables 1 and 4 we see that the length of the text is not strictly related to performance. More important is the style in which the text is written, which makes sense. Specifically, colloquial styles are more useful for finding

Table 5. Comparison of P@100 for original and extended Snowball golden standard

Text(s)	LDC	CCI	L'E	IPS	TLN	PPI1	PPI2	PPI3	PPI4	PPI5	PPI	HeG	AdA	N-T	All
Original	.53	.53	.62	.65	.62	.73	.71	.68	.71	.66	.72	.52	.41	.69	.72
Extended	.96	.70	.86	.90	.93	.90	.87	.85	.88	.86	.92	.70	.44	.89	.94

Table 6. Recall @100

	CCI	PPI1	PPI2	PPI3	PPI4	PPI5	PPI	IPS	L'E	LDC	HeG	TlN	AdA	NTT	All
NIDF	.03	.02	.03	.03	.02	.03	.02	.03	.03	.03	.05	.03	.02	.14	.14
TRS	.01	.01	.01	.01	.01	.01	.21	.01	.01	.01	.04	.01	.01	.16	.12
TF	.17	.25	.24	.24	.25	.23	.25	.21	.21	.19	.19	.22	.15	.25	.26
max	.48	.68	.69	.71	.69	.68	.82	.84	.68	.68	.43	.88	.41	.94	.95

(general) stopwords than technical ones. Indeed, the best performance is obtained on some volumes of PPI, which are not the longest documents but are written in a kind of journalistic style. Still quite high, but slightly lower, is the performance obtained on the longest single document, i.e., the stories of TlN. The two novels come immediately after, followed by the texts written using more particular styles, i.e., technical or poetry (plus HeG, which is narrative but is very short). Among the latter, precision on LDC (poetry) is slightly better than on CCI (technical), which might be partly unexpected, due to the old age and peculiar style of the former.

As expected, using many texts improves performance with respect to single texts, even if, differently from NIDF and TRS, performance on single texts is already very high for TF. While the improvement may not be outstanding compared to some single texts (e.g., TlN and PPI1), especially for the upper part of the ranking, a smoother decay in performance is clearly visible.

Based on the findings of [3], we also manually evaluated P@100 by extending the golden standard with some missing items. Indeed, [3] noted that many terms that we would safely consider stopwords are not in the Snowball stopword list, even if it does include other similar terms[8]. Results of TF using such an extended golden standard, shown in Table 5 in comparison with those obtained using the original Snowball golden standard, are even more impressive. Improvements are apparent and relevant on all documents except AdA (that, as noted, uses indeed very few stopwords). Especially interesting are the cases LDC (for the best increase in precision, by which it becomes the most effective document overall) and HeG (that also shows a gain of 0.18, raising the overall value to 0.70, albeit being a very short text, and thus not expected to contain very many different stopwords). A qualitative evaluation of the specific stopword lists returned by

[8] E.g.: 'essere', the infinitive form of verb 'to be', is missing, but many inflected form of that verb are in the list; 'fra' is not in the list, albeit being a very common alternate form of preposition 'tra', which is in the list; some modal verbs are in the list, but some others are not; etc.

the algorithm reveals that most often the wrong stopwords might be considered, however, domain-dependent stopwords (e.g., 'art.'—short for 'articolo', i.e. a law article—in CCI). This makes us confident in the possibility of building stopword lists also for domain-specific applications.

Finally, Table 6 compares the three approaches for recall (R@100). TF is again clearly superior, which was not obvious since it is well-known that increasing precision usually causes decreasing recall, and *vice versa*. Both NIDF and TRS never get close to it, not even on document collections. Again, TRS is better than NIDF only on aggregations of documents (except All), while on single document NIDF is better. Row *max* reports the portion of stopwords in the golden standard that actually occur in each (collection of) text(s). This is the maximum value that any possible approach, for any possible value of n may reach on those (collections of) texts. In practice, however, since the golden standard includes 279 stopwords, the maximum value @100 is actually bound by $100/279 = 0.36$. On many (collections of) texts, TF reaches values around 0.25, which we consider an outstanding result.

5 Conclusions and Future Work

Stopword removal is a fundamental pre-processing task for Information Retrieval applications, in order to improve effectiveness and efficiency of document indexing. It requires a list of stopwords, i.e. irrelevant terms to be identified and removed from the documents. As for other linguistic resources used in Natural Language Processing (NLP), stopword lists are language-specific. So, they might be unavailable for several languages, and manually building them is difficult. This paper focused on the automatic extraction of such stopword lists from texts written in a language, and studied the effectiveness of different approaches proposed in the literature, especially when applied to very small corpora (up to a dozen texts) and even to single texts.

Our hypothesis was that the simple TF technique, based on term frequency only (i.e., directly stemming from the definition of a stopword), might yield very good results, even in such extreme a setting. Not only our experiments confirmed the ability of TF to extract stopwords with quite good precision even from single, and even very short, texts (the shortest one used in the experiments included 890 words only). Very impressive was the fact that it outperformed two more complex techniques in the state-of-the-art, even in the case of small document collections. These techniques, designed for application to very large collections of documents, were both totally unable to learn significant or useful stopword lists when applied to single documents, in spite of the performance reported in the original paper. They were slightly better on small collections of documents, but still much worse than TF.

Given the good results on small corpora, a study of the behavior of TF on larger and more varied corpora should be carried out, to investigate up to what extent this technique still outperforms the other (more complex) techniques, and to understand under what conditions, if any, it is worth using either technique.

An indirect evaluation of performance through the effectiveness and efficiency of high-level NLP tasks based on the learned resources, as in [8], might be interesting. To make stopword identification fully automatic, another future work issue is to define an effective approach for distinguishing stopwords from non-stopwords among the candidate stopwords returned by the proposed technique. This is a complex task because of the irregular trend in the candidate ranking, providing little hints to determine a cutpoint. Indeed, the techniques proposed in the literature are not satisfactory.

References

1. Al-Shalabi, R., Kanaan, G., Jaam, J.M., Hasnah, A., Hilat, E.: Stop-word removal algorithm for Arabic language. In: Proceedings of the 2004 International Conference on Information and Communication Technologies: From Theory to Applications, pp. 545–549 (2004)
2. Cover, T.M., Thomas, J.A.: Elements of Information Theory. Wiley, Hoboken (1991)
3. Ferilli, S., Esposito, F.: On frequency-based approaches to learning stopwords and the reliability of existing resources – a study on Italian language. In: Serra, G., Tasso, C. (eds.) Digital Libraries and Multimedia Archives. IRCDL 2018, volume 806 of Communications in Computer and Information Science, pp. 69–80. Springer (2018)
4. Ferilli, S., Esposito, F., Grieco, D.: Automatic learning of linguistic resources for stopword removal and stemming from text. Procedia Comput. Sci. **38**, 116–123 (2014)
5. Fox, C.: A stop list for general text. SIGIR Forum **24**(1–2), 19–21 (1989)
6. Garg, U., Goyal, V.: Effect of stop word removal on document similarity for Hindi text. Eng. Sci. An Int. J. **2**, 3 (2014)
7. Kaur, J., Buttar, P.K.: A systematic review on stopword removal algorithms. Int. J. Future Revolut. Comput. Sci. Commun. Eng. **4**, 207–210 (2018)
8. Lo, R.T.-W., He, B., Ounis, I.: Automatically building a stopword list for an information retrieval system. In: Journal on Digital Information Management: Special Issue on the 5th Dutch-Belgian Information Retrieval Workshop, vol. 5, pp. 17–24 (2005)
9. Hans Peter Luhn: Keyword-in-context index for technical literature (kwic index). J. Assoc. Inf. Sci. Technol. **11**, 288–295 (1960)
10. Puri, R., Bedi, R.P.S., Goyal, V.: Automated stopwords identification in Punjabi documents. Eng. Sci. Int. J. **8**, 119–125 (2013)
11. Robertson, S.E., Sparck-Jones, K.: Relevance weighting of search terms. J. Am. Soc. Inf. Sci. **27**, 129–146 (1976)
12. Savoy, J.: A stemming procedure and stopword list for general French corpora. J. Assoc. Inf. Sci. Technol. **50**, 944–952 (1999)
13. Sinka, M.P., Corne, D.W.: Evolving better stoplists for document clustering and web intelligence, pp. 1015–1023. IOS Press, NLD (2003)
14. Sparck-Jones, K.: A statistical interpretation of term specificity and its application in retrieval. J. Doc. **28**, 11–21 (1972)
15. John Wilbur, W., Sirotkin, K.: The automatic identification of stop words. J. Inf. Sci. **18**(1), 45–55 (1992)

16. Xu, J., Bruce Croft, W.: Query expansion using local and global document analysis. In: Proceedings of the 19th Annual International ACM SIGIR Conference on Research and Development in Information Retrieval, pp. 4–11 (1996)
17. Zou, F., Wang, F.L., Deng, X., Han, S.: Evaluation of stop word lists in Chinese language. In: Proceedings of the 5th International Conference on Language Resources and Evaluation (LREC 2006), Genoa, Italy, May 2006, pp. 2504–2507. European Language Resources Association (ELRA) (2006)

BacAnalytics: A Tool to Support Secondary School Examination in France

Azim Roussanaly[1], Marharyta Aleksandrova[2(✉)], and Anne Boyer[1]

[1] University of Lorraine – LORIA, 54506 Vandoeuvre-lès-Nancy, France
{azim.roussanaly,anne.boyer}@loria.fr
[2] University of Luxembourg, 4365 Esch-sur-Alzette, Luxembourg
marharyta.aleksandrova@uni.lu, marharyta.aleksandrova@gmail.com

Abstract. Students who failed the final examination in the secondary school in France (known as *baccalauréat* or *baccalaureate*) can improve their scores by passing a remedial test. This test consists of two oral examinations in two subjects of the student's choice. Students announce their choice on the day of the remedial test. Additionally, the secondary education system in France is quite complex. There exist several types of baccalaureate consisting of various streams. Depending upon the stream students belong to, they have different subjects allowed to be taken during the remedial test and different coefficients associated with each of them. In this context, it becomes difficult to estimate the number of professors of each subject required for the examination. Thereby, the general practice of remedial test organization is to mobilize a large number of professors. In this paper, we present *BacAnalytics* – a tool that was developed to assist the rectorate of secondary schools with the organization of remedial tests for the baccalaureate. Given profiles of students and their choices of subjects for previous years, this tool builds a predictive model and estimates the number of required professors for the current year. In the paper, we present the architecture of the tool, analyze its performance, and describe its usage by the rectorate of the Academy of Nancy-Metz in Grand Est region of France in the years 2018 and 2019. *BacAnalytics* achieves almost 100% of prediction accuracy with approximately 25% of redundancy and was awarded a French national prize *Impulsions 2018*.

1 Introduction

Successful adoption of analytical tools in business and marketing impelled the usage of data analytics in education as well [1,9]. Data analytics usage in education defines 3 research directions: learning analytics (LA), educational data mining (EDM) and academic analytics [8,10]. Both LA and EDM aim to understand how students learn, with EDM having a primary focus on automated model discovery and LA having a stronger focus on keeping a human in the loop. The goal of academic analytics is to support institutional, operational and financial decision making. Academic analytics tools can be designed to assist in

M. Stettinger et al. (Eds.): ISMIS 2020, SCI 949, pp. 47–58, 2021.
https://doi.org/10.1007/978-3-030-67148-8_4

various tasks, such as identification of students at risk of failure [5,6], curriculum planning [7], organization of campus life services [4], building competency-based education courses [2] etc.

In this paper, we present a tool design to support the organization of the final examination in French secondary schools. French students who failed secondary school examination (known as *baccalauréat* or colloquially as *BAC*) are allowed to pass a remedial test. This test consists of oral examinations in two subjects of students' choice. Students announce their choices on the day of the remedial examination, which makes it impossible to calculate in advance the number of professors required to examine all students. Given the sensitive nature of the application, the general practice of academic rectorate is to mobilize a large number of professors. Our work presents *BacAnalytics* – a tool designed to estimate the required number of academic staff. To the best of our knowledge, this problem was not tackled in the literature before.

The rest of the paper is organized as follows. In Sect. 2, we describe the system of baccalauréat examination and the data and information that we used to construct the BacAnalytics tool. In Sect. 3, we present the architecture of the tool, its evaluation and impact. Finally, we conclude our work in Sect. 4.

2 Baccalauréat: Secondary School Examination in France

In this section we describe the organization of Baccalauréat in France. We also present the dataset provided to us by the Academy of Nancy-Metz that was used to build the predictive modules of BacAnalytics.

2.1 Baccalauréat organization

Secondary education in France is finalized with a baccalaureate examination. Unlike final examinations in secondary schools of other countries, BAC is not mandatory and it servers not for school completion, but for university entrance. There are 3 types of BAC: *baccalauréat général*[1] (general baccalaureate, BGN), *baccalauréat technologique*[2] (technological baccalaureate, BTN) and *baccalauréat professionnel*[3] (professional baccalaureate), see Table 1. Each type of BAC has multiple streams and many streams have multiple specializations. For example, stream STMG of BTN has 4 specializations: *GF* – finance management, *ME* – fast-moving consumer goods management, *RC* – communication and human resources, and *SI* – management information systems. Contrarily, stream ST2S has no specializations. Such a system allows providing specialized education for students with different needs and desires. For instance, professional baccalaureate is designed to prepare students for professional activities right after the school completion. At the same time, the vast majority of students sitting for

[1] https://eduscol.education.fr/cid46205/presentation-du-baccalaureat-general.html.
[2] https://eduscol.education.fr/cid46806/epreuves-du-baccalaureat-technologique.html.
[3] https://eduscol.education.fr/cid47640/le-baccalaureat-professionnel.html.

Table 1. Types of baccalaureate.

Type	Stream	Specialization[a]
General BGN	S – Scientific	3 6
	ES – Economics and Social Sciences	3
	L – Literature	9
Technological BTN	ST2S – Sciences and Technologies of Health Care	
	STI2D – Sciences and Technologies of Industry and Sust. Dev.	4
	STL – Sciences and Technology of Laboratory	2
	STMG – Sciences and Technologies of Management	4
	STD2A – Sciences and Technologies of Design and Applied Arts	
	STHR – Hospitality Industry and Business (before HOT)	
	TMD – Techniques of Danse and Music	
Professional baccalaureate: > 100 specialisations		> 100

[a] The first number corresponds to the number of defined specialization, and the second number corresponds to the number of specialized teaching programs. The latter differs in subjects and associated weights similar to specialization.

technological and general baccalaureate continue their studies as senior technicians (BTN) or at universities (BGN).

BAC consists of both oral and written exams in various subjects and each of them scores between 0 and 20. The set of subjects and associated weights depends on the type of BAC. Most of the subjects can be retaken during the remedial test. The list of subjects that can be retaken and their corresponding weights for BGN and some streams of BTN are presented in Table 2. Some of the subjects are the same for all students of the same stream (part of the table *general*), and others depend on the chosen specialization (part of the table *specialization*). The system of weights allows some subjects to be more important than others. Naturally, students usually study more for exams that carry heavier weights since the grade that they obtain in these exams have a bigger impact on their mean grade. The latter determines whether or not one passes the BAC. Students who average between 8 and 10 are permitted to sit for the remedial test (also called the 2d group as opposite to the initial 1st group examination). This is a supplementary oral examination, which is given in two subjects of the student's choice. Students announce their choices on the day of the examination, which does not allow calculating in advance the required number of professors. As shown in Fig. 1, the number of students going for the 2d group examination is quite high: around 15% of all students taking BAC examination for BGN and 17% for BTN. Also, we can notice a slight increasing trend from 2018 to 2019: +0.7% and +1.3% for BGN and BTN respectively. Thereby, the task of managing these students becomes more important.

Table 2. Subjects that can be taken in remedial examination and their weights.

General								Specialization		
Subject	Stream & weight							Stream	Subject & weight	
	S	ES	L	ST2S	STI2D	STL	STMG			
FRANCAIS**(7)**[a]	2	2	3	2	2	2	2			
HIST.GEOG.	3	5	4	2	2	2	2	S	ECO.AGR.TER.O	7/9
PHILOSOPHIE	3	4	7	2	2	2	2		SC. INGENIEUR	6/8
LANGUE VIV. 1	3	3	4/8[b]	2	2	2	3		SC. VIE TERRE	6/8
LANGUE VIV. 2	2	2	4/8	2	2	2	2	ES	SES - ECO APP	2
MATHEMAT.	7/9	5/7	.	3	4	4	3		SES - SC S PO	2
PHYS-CHIMIE	6/8	.	.	.	4	4	.	L	MATHEMAT.	4
SCIENCES	.	2	2	.	.	.	.		LCA LATIN	4
SC.ECO. SOC.**(7)**	.	7/9	.	.	.	.	.		LCA GREC	4
LITTERATURE	.	.	4	.	.	.	.		ARTS**(7)**	6
BIOL.PHYS.H.**(6)**	.	.	.	7	.	.	.	STL	BIOTECHNOL.	8
SC.TEC.SAN.S**(6)**	.	.	.	7	.	.	.		SPCL	8
SCI.PHYS.CH.**(6)**	.	.	.	3	.	.	.	STMG	GESTI.FINAN.**(6)**	12
ENS TECH TR**(4)**	.	.	.	.	8	.	.		MERCATIQUE**(6)**	12
ECO.-DROIT	.	.	.	.	.	.	5		RH.COMMUN.**(6)**	12
MANAG.ORGAN.	.	.	.	.	.	.	5		SYST.INFO.G.**(6)**	12

[a] Number of students a professor can examine during a one-day session (if not specified, the number of students is equal to 9).
[b] Weight depends on the chosen specialization.

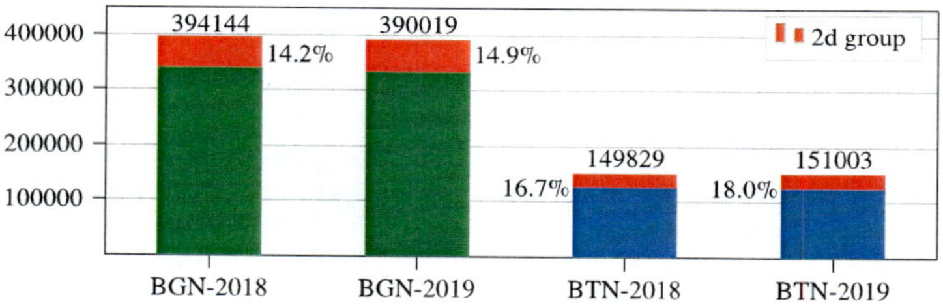

Fig. 1. Number of students registered for BAC in France for BGN and BTN per year. This data was provided by the Academy of Nancy-Metz.

2.2 Dataset

In order to overcome this problem, we collaborated with the Academy of Nancy-Metz[4] of the French region Grand Est. They provided us with the anonymized historical information about students who took the remedial test of baccalauréat for BGN and some streams of BTN.[5] The streams of BTN that were not considered in this work due to the lack of data are presented in italics in Table 1.

[4] http://www.ac-nancy-metz.fr/.

[5] In this work, we use this anonymized dataset with respect to article 6 clause (f) of GDPR: "for the purposes to the legitimate interests pursued by the controller", https://gdpr-info.eu/art-6-gdpr/.

The distribution of the number of students of the 2-d group from our dataset by years and streams of BAC is presented in Fig. 2 for BGN and Fig. 3 for BTN. The collaboration started in 2017 and in 2018 and 2019 BacAnalytics was tested in field conditions. In this paper, we discuss in details the tool's performance for 2018. Additionally, we present the final results from 2019 to support our closing conclusion.

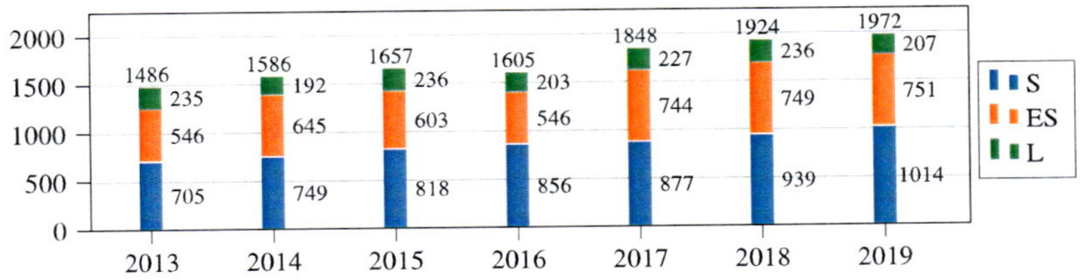

Fig. 2. BGN: number of students of the 2-d group per year.

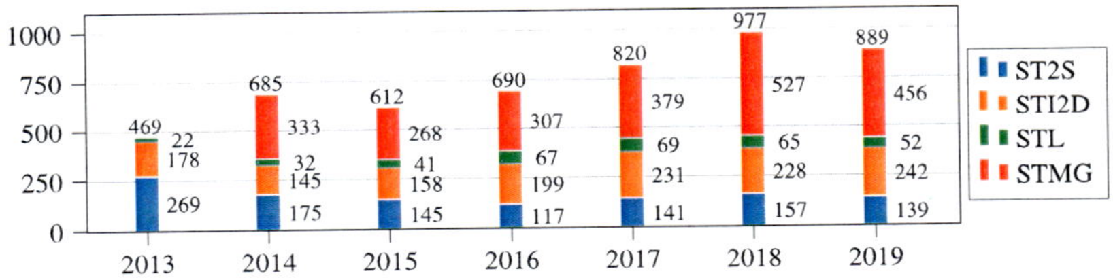

Fig. 3. BTN: number of students of the 2-d group per year.

The provided dataset contains general information about students (Table 3) and their performance in the 1-st group examination (Table 4). From Table 3, for every student, we know his/her unique id, examination year, type, stream and specialization of baccalaureate, associated geographical center for the remedial test, the 2 subjects chosen for the remedial test (choice#1 and choice#2) and the corresponding grades. Additionally, we know how well every student performed in the 1-st group examination, e.g. we know what grades he/she got for every subject, see Table 4.

BacAnalytics tool was developed based on this dataset and the general information about the organization of BAC presented above. Using the historical data, our tool aims to predict the 2 subjects chosen by a student, that is the values of choice#1 and choice#2 from the Table 3.

Table 3. Example of the data with information about students and their performance in the 2d group test.

id	year	type	stream	specialization	center	choice#1	grade#1	choice#2	grade#2
1	2013	BTN	ST2S		054G	subject_3	11	subject_1	12
2	2013	BTN	STMG	ME	057Z	subject_1	13	subject_7	9
...	...	...	...	...	...	...	...	...	...
127	2015	BGN	ES		088G	subject_2	8	subject_5	9
...	...	...	...	...	...	...	...	...	...

Table 4. Example of the data with information about performance of the students in the 1st group test. Note that the number of subjects depends on BAC type, stream, and specialization and can be different for different students.

id	subject	grade
1	subject_1	14
1	subject_2	9
1	subject_3	16
...	...	...
127	subject_1	15
...	...	...

3 BacAnalytics Tool

In this section, we describe the architecture of BacAnalytics, discuss its evaluation and impact.

3.1 Architecture

The general architecture of the BacAnalytics tool is presented in Fig. 4. We start with a construction of a model for predicting students' choices. In this work we used WEKA implementation of the *Random Forest* classifier with default parametes [3] as a predictive model, however, the usage of other algorithms is also possible. To build the predictive model for the year m, we use historical information for previous years $m-1, m-2, \ldots$ as a training dataset. In particular, we train a classifier to predict choice#1 and choice#2 from Table 3 using the type and stream of BAC (Table 3) and students' performance in the 1st group examination (Table 4). Next, using corresponding information about students for the year m, we predict their choices. To obtain predictions of multiple choices, we select top-N most probable subjects according to the model's output. In this paper, we perform evaluations for $N = 2$ and $N = 3$. We refer to this stage as **Step 1** prediction. An example of Step 1 prediction for $N = 2$ is presented in Table 5.

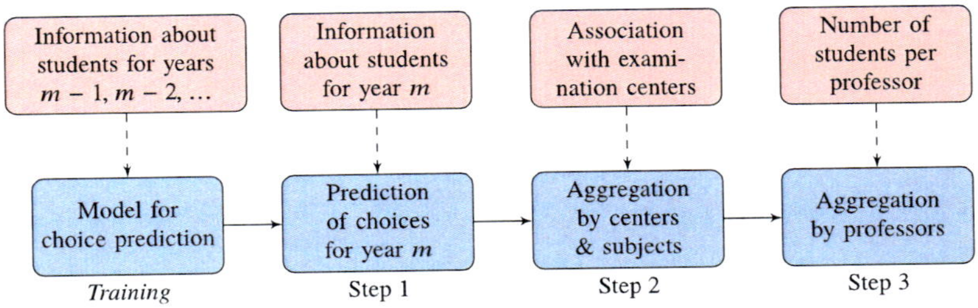

Fig. 4. BacAnalytics architecture.

Table 5. Example of step 1 prediction

id	type	stream	center	choice#1	choice#2
1	BTN	STMG	054G	FRANCAIS	**MERCATIQUE**
2	BTN	STMG	054G	**MERCATIQUE**	*GESTI.FINAN.*
3	BTN	STMG	054G	SYST.INFO.G	**MERCATIQUE**
4	BTN	STMG	054G	SYST.INFO.G	**MERCATIQUE**
5	BTN	STMG	054G	**MERCATIQUE**	*GESTI.FINAN.*
6	BTN	STMG	054G	*GESTI.FINAN.*	**MERCATIQUE**
7	BTN	STMG	057Z	SYST.INFO.G	*GESTI.FINAN.*
8	BTN	STMG	057Z	*GESTI.FINAN.*	SYST.INFO.G
9	BTN	STMG	057Z	**MERCATIQUE**	SYST.INFO.G
10	BTN	STMG	057Z	SYST.INFO.G	*GESTI.FINAN.*
11	BTN	STMG	057Z	*GESTI.FINAN.*	SYST.INFO.G
12	BTN	STMG	057Z	**MERCATIQUE**	SYST.INFO.G

On **Step 2**, the students are distributed between relevant examination centers and their choices are aggregated by subjects. In the prediction example presented in Table 5, there are 2 examination centers and students are predicted to choose 2 of the following 4 subjects: GESTI.FINAN, SYST.INFO.G., MERCATIQUE and FRANCAIS. The corresponding aggregation results are presented in Table 6. These results are obtained by counting the number a particular subject was predicted in every examination center.

Finally, on **Step 3** we perform final aggregation to estimate the required number of professors. For this, we divide the number of times every subject was predicted to be chosen by the number of students a professor can examine during a one-day session, see note a for Table 2. If a resulting number is not an integer, we round it using the *ceiling* function. From Table 2 we can see that one professor of MERCATIQUE can examine 6 students. Thereby, the predicted number of professors for center 054G will be $ceiling(6/6) = 1$. For center 057Z the value will be $ceiling(2/6) = 1$. Finally, we also ensure that there is at least

Table 6. Example of step 2 prediction

Center	GESTI.FINAN	MERCATIQUE	SYST.INFO.G	FRANCAIS
054G	3	6	2	1
057Z	4	2	6	0

one professor of every subject. In this way, for the considered example 1 professor of FRANCAIS will be predicted for both centers, even thought no students were predicted to choose this subject in the examination center 057Z.

Step 1, step 2, and step 3 produce predictions that can be compared with the corresponding real values. Although only the output of step 3 is required for the organization of the remedial test, in the next section we evaluate the performance of BacAnalytics on all 3 steps. This allows us to understand better the performance of the tool and identify possible ways of improvement.

3.2 Tool Evaluation

To evaluate the performance of the predictive models, we use *accuracy* (acc) as the main metric. The value of accuracy is defined as a fraction of the number of correctly predicted instances ($\#corr$) to the total number of instance to be predicted ($\#to\_predict$), see Eq. (1). Additionally, as in our evaluation we use $N > 2$, some of the predictions will be redundant. We evaluate *redundancy* (red) of a model according to the formula in Eq. (2).

$$acc = \frac{\#corr}{\#to\_predict} \quad (1) \qquad red = \frac{\#predicted - \#corr}{\#predicted} \quad (2)$$

Step 1: per-student prediction. We evaluate the performance of predictions on step 1 in two ways: how accurately the model can predict at least 1 subject of the student's choice ($c = 1$) and both of them ($c = 2$). Corresponding results for 2018 are presented in Fig. 5 and Fig. 6. We can see that even when taking only 2 predicted subjects ($N = 2$), the model is quite accurate in $c = 1$ scenario, with the lowest accuracy being 0.93 for STMG stream of BTN. However, the performance drops significantly if we require both choices to be predicted correctly ($c = 2$). The lowest accuracy, in this case, is 0.28 for STL stream of BTN and the highest value of accuracy for all streams does not exceed 0.56. For $N = 3$ overall performance increases. Accuracy of prediction of at least 1 subject comes very near to 1 for most of the streams, except STMG with corresponding value of 0.96. Accuracy for $c = 2$ also increases, however, for streams STL and STMG it is below 0.6. Thereby, we can conclude that the model stays unable to reliably predict both choices of students. Generally, prediction model performs better for BGN with ES stream having the lowest prediction accuracy. As for BTN, the predictive model struggles the most with STL and STMG streams.

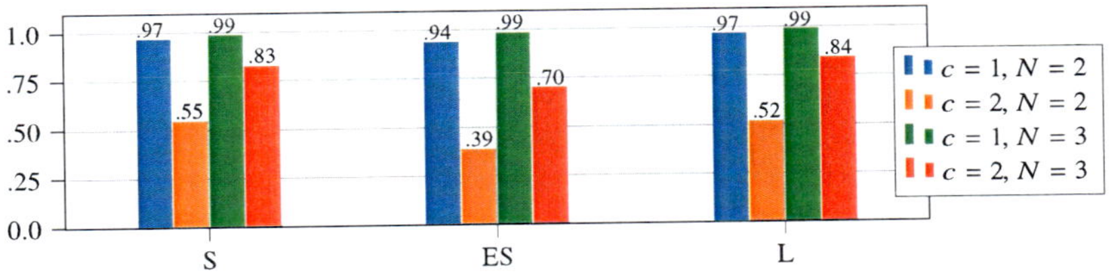

Fig. 5. BGN: accuracy of step 1 prediction for 2018.

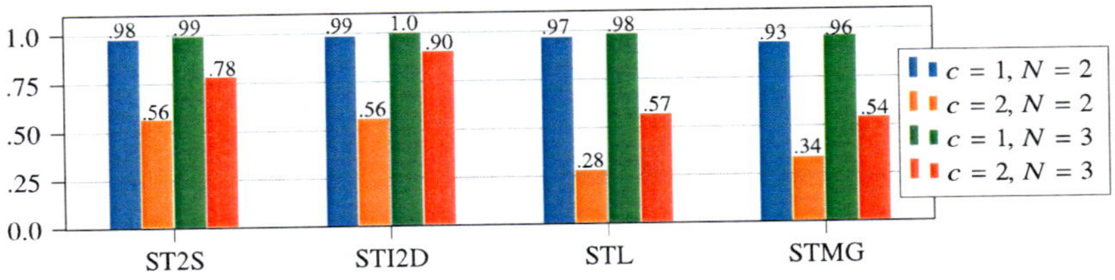

Fig. 6. BTN: accuracy of step 1 prediction for 2018.

Step 2: per-subject prediction. To evaluate the results of step 2 prediction, we use both accuracy and redundancy. The results for 2018 are presented in Fig. 7 and Fig. 8. We can notice a considerable improvement in performance as compared to the results of step 1 for prediction of 2 choices ($c = 2$). $N = 2$ scenario results on average in $acc = 0.89$ and $red = 0.11$. For $N = 3$, we can predict the number every subject is chosen with an average accuracy of 0.98. This also comes at a cost of increased redundancy with the average value of 0.34. On the opposite, the corresponding value of accuracy for $c = 2, N = 3$ scenario of step 1 prediction is only 0.74. Such improvement is explained by the fact that when performing aggregation per subject, some incorrect predictions can compensate each other. For example, if in the same examination center student i chose mathematics and French and student j chose geography and philosophy, then predicting mathematics and philosophy for student i and geography and French for student j results in absolutely correct per-subject prediction.

Step 3: per-professor prediction. The most important and practically useful indicator of BacAnalytics performance is the evaluation of predictions on step 3. The corresponding results for 2018 are given in Fig. 9 and Fig. 10. We can see further performance improvement. For instance, for $N = 2$ the average accuracy is equal to 0.95 and the average redundancy to 0.08. When we increase the value of N to 3, BacAnalytics can correctly predict the number of required professors for all streams but STMG of technological baccalaureate with the corresponding value of accuracy being 0.96. The average value of redundancy, in this case, is 0.25 which is a significant reduction as compared to $red = 0.35$ for step 2 prediction with $N = 3$. The reason for this improvement is error compensation due to further aggregation. For example, if the French language was chosen by

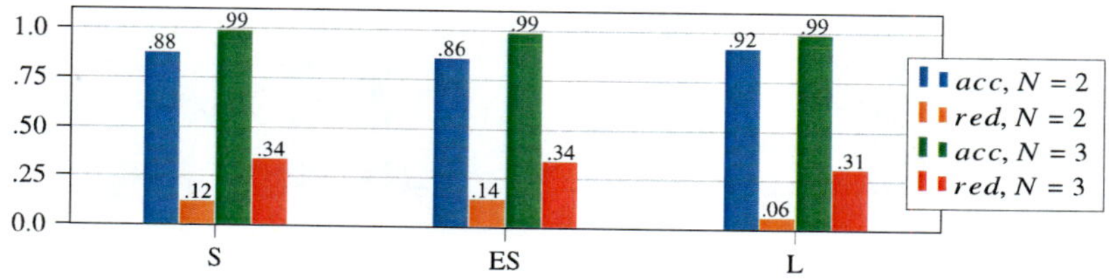

Fig. 7. BGN: accuracy and redundancy of step 2 prediction for 2018.

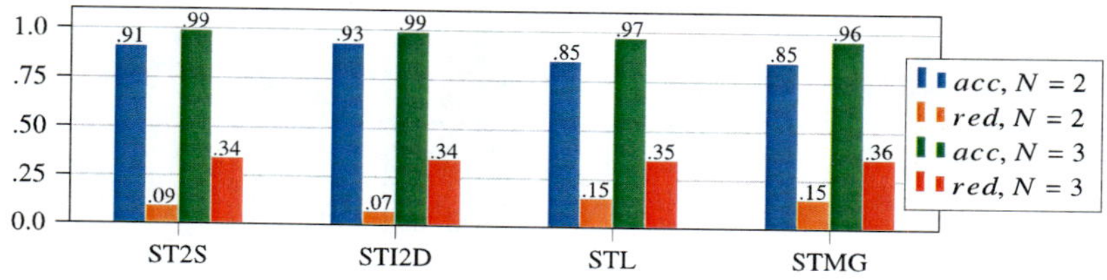

Fig. 8. BTN: accuracy and redundancy of step 2 prediction for 2018.

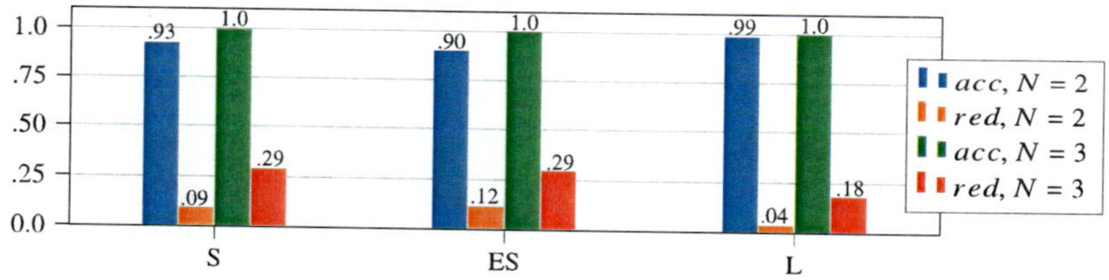

Fig. 9. BGN: accuracy and redundancy of step 3 prediction for 2018.

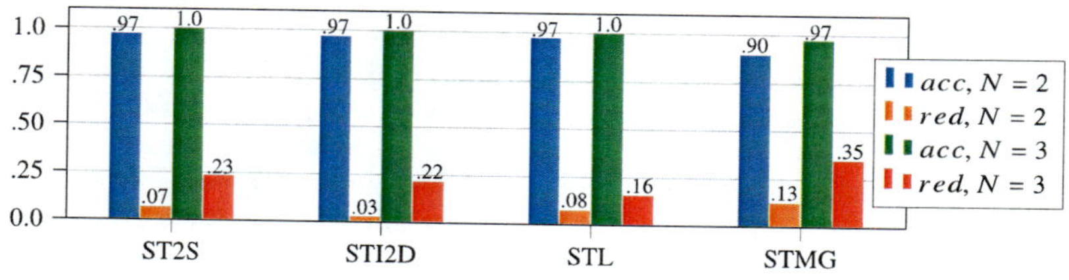

Fig. 10. BTN: accuracy and redundancy of step 3 prediction for 2018.

12 students in reality and BacAnalytics estimated this value to be 8, then per-professor aggregation will result in the same number of required professors that is equal to 2, as 1 professor of French can examine 7 students, see note a for Table 2. Overall, we can conclude that the results obtained by BacAnalytics with $N = 3$ are good enough to be used in practice. This statement is also supported by the final results obtained for the year 2019 with only a minor decrease in accuracy, see Fig. 11.

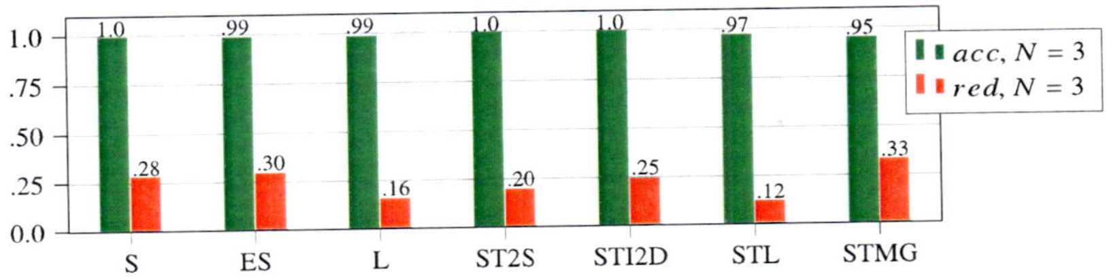

Fig. 11. Accuracy and redundancy of step 3 prediction for 2019, $N = 3$.

3.3 Impact of BacAnalytics

As mentioned before, BacAnalytics was developed in collaboration with the Academy of Nancy-Metz in Grand Est region of France. After preliminary model evaluation on the data for years 2013-2017, the tool was used during the preparation of the remedial test in 2018 and 2019. Given the nature of the application, the rectorate of the academy decided to use $N = 3$ as a default value. As was shown above, such settings resulted in quite accurate predictions. In general, both professors and administration reported a more peaceful and enjoyable experience. BacAnalytics was also awarded the national French prize Impulsion 2018 in nomination "Innovation"[6],[7].

4 Conclusions and Future Work

This paper presents BacAnalytics – a tool that was developed to assist in the preparation of the remedial test for the French secondary school examination called baccalauréat. The baccalauréat system allows the students to announce the subjects to be retaken on the day of the remedial test. Given that all examinations on this day are oral, a large number of professors have to be mobilized to fulfill the possible demand. BacAnalytics utilizes historical information about students' choices to estimate the number of required professors. It achieves

[6] http://www.ac-nancy-metz.fr/prix-impulsions-2018-l-8217-academie-de-nancy-metz-primee-au-niveau-national--120140.kjsp.

[7] https://www.education.gouv.fr/cid136476/trois-laureats-primes-au-prix-impulsions-2018-de-la-modernisation-participativeprixduprojetinnovant.html.

almost 100% prediction accuracy at the price of approximately 25% redundancy. This tool was successfully employed by the Academy of Nancy-Metz in the years 2018 and 2019 and was awarded a French national prize in nomination "Innovation".

The evaluation results presented in the paper show, however, that the tool can be improved. Some of the baccalauréat series are more difficult to prediction than others. In future work, we consider constructing more refined models for these cases. Additionally, we want to use the approaches of error-aware data mining to incorporate the feedback and further improve the results.

References

1. Baepler, P., Murdoch, C.J.: Academic analytics and data mining in higher education. Int. J. Scholarsh. Teach. Learn. **4**(2), 1–9 (2010)
2. Durand, G., Goutte, C., Belacel, N., Bouslimani, Y., Léger, S.: A diagnostic tool for competency-based program engineering. In: Proceedings of the 8th International Conference on Learning Analytics and Knowledge, pp. 315–319 (2018)
3. Eibe, F., Hall, M.A., Witten, I.H.: The weka workbench online appendix for "data mining: practical machine learning tools and techniques". Morgan Kaufmann (2016). https://www.cs.waikato.ac.nz/ml/weka/Witten_et_al_2016_appendix.pdf
4. Heo, J., Lim, H., Yun, S.B., Ju, S., Park, S., Lee, R.: Descriptive and predictive modeling of student achievement, satisfaction, and mental health for data-driven smart connected campus life service. In: Proceedings of the 9th International Conference on Learning Analytics & Knowledge, pp. 531–538 (2019)
5. Lauría, E.J., Moody, E.W., Jayaprakash, S.M., Jonnalagadda, N., Baron, J.D.: Open academic analytics initiative: initial research findings. In: Proceedings of the Third International Conference on Learning Analytics and Knowledge, pp. 150–154 (2013)
6. Lawson, C., Beer, C., Rossi, D., Moore, T., Fleming, J.: Identification of 'at risk' students using learning analytics: the ethical dilemmas of intervention strategies in a higher education institution. Educ. Tech. Res. Dev. **64**(5), 957–968 (2016)
7. Morsy, S., Karypis, G.: A study on curriculum planning and its relationship with graduation GPA and time to degree. In: Proceedings of the 9th International Conference on Learning Analytics & Knowledge, pp. 26–35 (2019)
8. Romero-Zaldivar, V.A., Pardo, A., Burgos, D., Kloos, C.D.: Monitoring student progress using virtual appliances: a case study. Comput. Educ. **58**(4), 1058–1067 (2012)
9. Van Barneveld, A., Arnold, K.E., Campbell, J.P.: Analytics in higher education: establishing a common language. EDUCAUSE Learn. Initiat. **1**(1), l–ll (2012)
10. Viberg, O., Hatakka, M., Bälter, O., Mavroudi, A.: The current landscape of learning analytics in higher education. Comput. Hum. Behav. **89**, 98–110 (2018)

Towards Visual Concept Learning and Reasoning: On Insights into Representative Approaches

Anna Saranti[1(✉)], Simon Streit[1], Heimo Müller[1], Deepika Singh[1], and Andreas Holzinger[1,2]

[1] Medical University Graz, Auenbruggerplatz 2, 8036 Graz, Austria
{anna.saranti,simon.streit,Heimo.Muller,deepika.singh, andreas.holzinger}@medunigraz.at

[2] xAI Lab, Alberta Machine Intelligence Institute, Edmonton T6G 2H1, Canada

Abstract. The study of visual concept learning methodologies has been developed over the last years, becoming the state-of-the art research that challenges the reasoning capabilities of deep learning methods. In this paper we discuss the evolution of those methods, starting from the captioning approaches that prepared the transition to current cutting-edge visual question answering systems. The emergence of specially designed datasets, distilled from visual complexity, but with properties and divisions that challenge abstract reasoning and generalization capabilities, encourages the development of AI systems that will support them by design. Explainability of the decision making process of AI systems, either built-in or as a by-product of the acquired reasoning capabilities, underpins the understanding of those systems robustness, their underlying logic and their improvement potential.

Keywords: Artificial Intelligence · Human intelligence · Intelligence testing · IQ-test · Explainable AI · Interpretable machine learning · Representation learning · Visual concept learning · Neuro-symbolic computing

1 Introduction and Motivation

Despite the enormous progress in Artificial Intelligence (AI) and Machine Learning (ML) along with the increasing amount of datasets and computer performance, Yoshua Bengio in his NeurIPS 2019 Posner Lecture on December, 11, 2019, emphasized that we are still far from achieving human-level AI and even children can perform some tasks better than the best machine learning models [29]. For a better understanding of these challenges we follow the notion of Daniel Kahnemann, who described two systems of (human) cognition in his famous book [21]. Such an approach has been implemented by Anthony, T. [2] with a reinforcement learning algorithm which decomposes the problem into separate planning and generalisation tasks to play the board game Hex.

M. Stettinger et al. (Eds.): ISMIS 2020, SCI 949, pp. 59–68, 2021.
https://doi.org/10.1007/978-3-030-67148-8_5

System 1 is working intuitively, fast, automatic, and **unconscious** (e.g. determine that an object A is at a greater distance than object B)

System 2 is working slow, logical, sequential and **conscious** (e.g. count the number of objects in a certain area).

Bengio pointed out that the most important future work will be to move towards deep learning models that can not only just operate on vectors, but also operate on sets of objects. This is an important motivation for research that focuses on symbolic methods, that already reason over individual objects, their properties, their relations as well as object sets [28]. One of the possible approaches for the combination of the two systems would be the implementation of symbolic logic by specific neural network architectures [4,36], combined with a description of the concepts by a domain specific language (DSL) [6]. The resulting executable program consists of the explanation that provides traceability of the learned logic and decision mechanisms and can be processed independently [26]. Even if explainability has a negative impact on performance, it supports the uncovering of biases, adversarial perturbations and deficiencies, thereby enhancing the robustness of AI systems through transparency [3].

The remainder of this paper is structured as follows: In Sect. 2, a selection of representative benchmark datasets for reasoning are analyzed; their requirements, basic characteristics and evolution is described. In Sect. 3, many representative non-symbolic approaches that tackle visual reasoning with different methodologies are explained. The transition to symbolic methods is presented in Sect. 4, along with the benefits, reasoning and generalization capabilities of those methods. A specific dataset, the Kandinsky patterns, that is of importance for visual reasoning in the medical domain and applies to visual concepts considering object sets and the visual scene as a whole, is presented in Sect. 5. Conclusion and future work summarizes all the sections, and maps out the setup for the development of the approach to be followed for this particular dataset.

2 Benchmark Datasets for Reasoning

Image classification, object localization, and semantic-to-instance segmentation applications using deep learning methods were already developed, by the time Microsoft introduced the COCO (MS-COCO) dataset [5,25]. Research in the field of semantic description extraction from a scene in the form of captions has been systematized and evolved since then. Captions are a textual descriptions about the content of an image and there can be a lot of different ones for one particular image. Machine learning algorithms can regard the captions as weak labels [22]; instead of providing a hard label for each image indicating f.e. the cancer type, a set of textual descriptions help the algorithm to compute the correlation between a word in a sentence and the corresponding image part it refers to, which is called visual grounding. The work of Karpathy [22] is representative for the generation of textual descriptions for regions of an image associated with them, which is achieved by the alignment of visual and textual embeddings in a common vector space. The image is processed by a convolutional neural network and the text by a bidirectional recurrent network. As a result, a region in the image corresponds to a sequence of words that describe it [12].

In contrast to the previous image processing benchmarks, the creators of MS-COCO recognized the necessity of the context in the recognition of objects that are relatively small or partially occluded; this is further highlighted by more recent research works [24]. Therefore, the images that are selected contain the objects in their natural environment and in non-iconic views. Because of the noise and variability in the images, the models do not necessarily perform better for instance detection tasks, but have increased generalization capabilities according to a cross-dataset generalization metric [35].

Research evolved from captioning to question answering (QA) systems for a variety of reasons. First of all, a question answering scheme is considered more natural than labelling or captioning. Furthermore, the reasoning models use the information contained in the question, for correctly answering the question [28]. The desired output after the training process is the correct answering of questions of various complexity levels. Each question tackles different concepts and reasoning challenges.

A representative dataset in the field for visual question answering systems (VQA) is CLEVR [18]. The dataset consists of images containing a constrained set of objects with predefined variability of attributes in different recognizable constellations. The ground truth is known by construction and the associated questions to each image is generated by a functional program composed by a chain or tree of the reasoning steps, that the machine learning algorithm under test will have to possess, to be able to answer the question correctly. It can be used for visual question answering systems to uncover shortcomings of machine learning algorithms that although performant, base their decision on statistical correlations and can generalize only to a limited extend to unseen data.

The text in the question plays also a role in overcoming biases; a uniform answer distribution must be ensured through rejection sampling. Longer questions tend to need longer reasoning paths and are considered more complex. Nevertheless, the researchers observed cases where the reasoning steps were not all correctly followed, but the answer was still correctly answered. The fact that state-of-the-art deep learning models that combined image processing and textual aligning were concentrating in absolute object positions and could not adapt to constellations where only an attribute value was changed, indicated that the attribute representations are not disentangled from the objects. This leaves less potential to generalization on unseen scenes, even if they are just combinations of known objects characteristics.

Reasoning datasets exercising the temporal and causal reasoning capabilities of deep learning models are currently being researched. CLEVERER [37] extends the CLEVR dataset with the introduction of collision events between the objects, thereby motivating the deep learning model having additional predictive and counterfactual capabilities. The recognition of objects, their dynamics and relations as well as the events, is supported by motion. It is crucial for causal relations recognition to have a separate object-centric representation component (supervsion) in the model, which is overtaken by a neural network. Causal reasoning in contrast, is expected to be tackled by a symbolic logic component, supported by the input questions in the form of an implementable program.

Basic generalization capabilities are measured by the ability of the model to retain a good performance for different dataset training/test split, where some attribute or property is changed - often in an opposite way. Ideally, there should be no performance degradation across dataset splits. All the aforementioned benchmarking datasets were designed and implemented with specific rules that addressed the problem of bias. For example in CLEVERER [37], every posed question must have a balanced amount of images for each possible answer. Appearance variability [25] needs to be ensured by various methods including filtering out iconic images, selection of independent photographers, particularly if the images are gathered from Internet resources [22]. Captions must fulfill particular statistics, need to be evaluated w.r.t. the agreement degree of different persons [5,25]. Existing datasets evolve and improve regularly, and new ones are being continuously created.

3 Non-symbolic Reasoning and Representation Learning Methods

Probabilistic graphical models [34] are used in interpretable description of images. Scene graphs are conditional random fields that are used to model objects, their attributes and relationships within an image, through random variables connected with edges [20]. The visual grounding does not apply on a specific part of the image that corresponds to a word, but on the scene graph with respect to the likelihood of the image as a whole. They can be learned from a set of images, and also retrospectively used to generate images from the learned models [17] with the use of a graph neural network that processes the graphical model by graph traversal. The results of the research showed that their generalization capabilities does not include rare elements. Since the graphical model's random variables are constructed from components of images that are already seen, the performance on a test set with a valid but unseen configuration at training time, is not satisfactory. On the contrary, the interpretability of this approach is built-in by the graphical model, since each node and relation is understood.

Another approach describes the contents of a static image containing objects of various visual characteristics and perceivable groupings by a scene program [26]. This representation enables the discovery of regularities such as symmetry and repetition that can be achieved with the definition of a domain-specific language (DSL). The DSL defines the grammar necessary for program generation, which bases firstly on object identification and attribute prediction with a Mask R-CNN [10] and a ResNEt-34 [11] and secondly on a sequence to sequence (seq2seq) LSTM model that outputs the next token of the scene program. Although the DSL grammar comprises the human prior and has constrained and pre-specified expressions, the generated program can theoretically have an arbitrary length to support generalization. The datasets that are used for testing use a model that is pretrained on the synthetic scenes training dataset, after being preprocessed by the a corresponding Mask R-CNN.

The benefits of this approach include correct scene programs even in the case of partially hidden objects, the ability to generate many different programs from the same input image and the manipulation of the program for the generation of new images that are perceived as realistic. The generalization capabilities encompass the correct recognition of groups of objects in scenes with randomly placed synthetic objects and in scenes comprised by Lego parts, with a better performance from the baseline method. The number of object groups recognized in the test set generalizes to one more than the maximum number of groups encountered in the training set. When a valid scene program is subjected to minor changes, the generated real images of Lego parts ordered in a grid have also better L2 distance performance than a corresponding autoencoder.

Other approaches use a human domain-expert that defines a knowledge base (KB) containing the ontologies and rules relevant for modelling. Those are interpretable, assumed to be correct, and the reasoning can be overtaken by specially designed neural network architectures [3,27,32]. Relational Networks [33] deal with relational questions between all pairs of objects in a scene, but the learning of concepts involving sets of objects in a scene as a whole, or metaconcepts is not supported.

4 Symbolic Reasoning and Representation Learning Methods

Since the benchmark datasets described in Sect. 2 challenged the reasoning abilities of artificial intelligence methods, the idea of modelling those reasoning processes and the concepts involved in them with neural modules [1] goes back to the work of Johnson [19]. A program implementing all the reasoning steps is generated by an LSTM that takes the image and question as input. Those models did not yet learn disentangled representations of attributes like colors and shapes, needed an explicit module for every concept necessary to answer the question, and did not show robustness in the face of novel questions. Neural module networks have recently shown their capabilities in text-only reasoning tasks [7]. Probabilistic scene graphs became able to achieve disentangled representations also with the use of neural state machines [16], which reason sequentially over the constructed graph.

Symbolic AI approaches have been shown to achieve a degree of generalization and disentanglement of representations. In the work of [38], the scene is parsed first to get its structure containing features like size, shape, color material and coordinates - this procedure is also called inverse graphics. It disentangles the representation of the scene from the symbolic execution engine that follows it, thereby gives the ability for generalization since other types of scenes can be processed by the same model, as long as their representation is analogous.

The question is used as an input to an seq-2-seq bidirectional LSTM that produces a series of Python modules as output which will process the structural representation of the image. The possible modules and their functionality are predefined; by that means the human prior is contained in the solution. There is a correspondence between each logical operator and the reasoning ability that is presupposed for answering the question in a right way. The questions consider object attributes, properties as well as counting under specific constraints.

The benefits of this type of disentanglement is the avoidance of over-fitting at tested dataset splits, the low memory requirements and the increased performance in comparison to several methods. All components of the model are considered interpretable; from the scene representation, the input question expressed in natural language, the answer as well as the generated program. The generalization capabilities are exercised on a Minecraft scenes dataset that has a slightly larger structural representation than the CLEVR dataset. The generalization to completely unseen natural scenes that are much more complex than the ones used at training time, is not yet achieved by this method.

The drawback of the [38] is the high grade of supervision due to the predefined scene representations and programs. To establish a trainable representation for both the visual and language features the authors of [28] introduced a novel method of optimization. The key change is the replacement of constant scene representation with a visual-semantic space, allowing for a variable number of different visual concepts, attributes and relations. This is achieved by assigning neural operators and specific embedding vectors towards the output objects, as analyzed by the Mask-RCNN. The neural operators are simple linear layers of neurons, while the concept embeddings are initialized randomly. The learning is achieved by using curriculum learning that separates concepts first (e.g. shape, color from spacial relations) and gradually increases in difficulty as the semantic space is consolidated. In contrast to [38], the semantic parser processes the questions without using annotations on programs at all. It rather directly generates candidate programs to be evaluated, while the optimization is then guided by REINFORCE directly. As in aforementioned approaches, the semantic parsing concepts are identified by handcrafted rules; the questions follow certain templates in order for object level concepts (e.g. shape), attributes (e.g. color) or spatial relations to be identified correctly. This also ensures that the neural operators and relating embedding vectors can be learned correctly using the DSL program implementations. Those design decisions achieved generalization capabilities that extend to more objects, different attribute combinations and zero-shot learning of a color.

In this work, the interpretability is a by-product of the program that is internally constructed and parsed to answer the posed question. The sequential steps describe the operations applied to the input image (w.r.t. the question) and ensure re-traceability. The specification of a domain-specific language (DSL) reinforces the explainability aspect, since it predefines the concepts and their mapping to the program operations.

Disentangled representations can also be achieved through metaconcepts [9], which consist of abstract relations between concepts. The image preprocessing stages are similar to the ones presented in [38] and [28], but the further implementation implements each metaconcept (for example: synonym) through a symbolic program and the concepts, as well as the object representations, as vector embeddings. The neural operator takes the concept embeddings as inputs and performs a classification, deciding if two concepts have a metaconcept relation. This architecture requires a questions dataset that is enhanced with questions considering metaconcepts, but the performance of visual grounding of particular concepts is more data efficient and does not be supported by a large number of examples containing them.

5 Kandinsky Patterns Dataset

The Kandinsky patterns dataset is an exploration environment for the study of Explainable AI that is designed to address learning and generalization of concepts in the 2D medical images domain [14]. The necessity of explanation in this domain encourages the use of textual descriptions of medical images, as well as more interactive question and answering systems. Heatmapping approaches like Layer-wise Relevance Propagation (LRP) [8] provide insights in the decision-making process of a deep learning system, but this cannot replace a diagnosis. On the other hand, captions of those images expressing predefined relevant concepts, do not necessarily uncover the operation, causes or decision criteria of a decision of a deep learning algorithm [3].

Most of the aforemenetioned datasets contain rendered 3D objects, taking into consideration the camera position in such a way that bias is avoided. A 2D dataset for multimodal language understanding containing objects without overlapping and has similarities the Kandinsky patterns is ShapeWorld [23]. The motivation for the creation of this dataset was also to uncover biases and encourage machine learning algorithms to exploit the combination of concepts. They differ from the Kandinsky patterns in the way the captions are generated, which in this case is synthetic and according to the rules of a prespecified grammar. The mapping of entities to nouns, attributes to adjectives and relations between the entities to verbs support the correspondence between f.e. color-describing attributes or positional comparison expressions like "left of" ensures the systematization of the caption generation.

The specifications of the experiments ensure that the evaluation dataset has different characteristics from the training dataset. A deep learning algorithm that achieves that goal, is considered to perform some kind of zero-shot learning. Overall the authors consider this dataset as a unit test for multimodal systems, since the combination of concepts is necessary to achieve the goals at evaluation time. The tested evaluation multimodal deep neural network architectures, comprised usually by a CNN and LSTM module, perform object recognition tasks with near 100% accuracy while having near-random performance (50%) in spatial relations classification. These experiments are in accordance with the ones made with the CLEVR dataset and underline the necessity of specifically designed reasoning models.

The concepts that are relevant for the medical domain go beyond object attributes like color, shape or relations between pairs of objects; they refer to properties of the whole object set in the scene like symmetry, arithmetic relations and further specific constellations [31]. Those more complex concepts will need an extended domain language that bases on the effective reasoning of the simpler concepts - as learned for datasets like CLEVR - as well as meta-concept learning, which consists of relations between concepts. For example, an arithmetic relation between different objects in a scene, presupposes the ability of the model to count, which is a simpler object-level concept.

6 Conclusion and Future Work

Since the Kandinsky patterns dataset will address reasoning over more complex concepts and metaconcepts, it should be first extended by a corresponding set of questions, following the principles of the benchmarking datatasets described above. This can be the first step for future research on dialogue systems [30] for future AI-interfaces. Furthermore, the use of executable symbolic programs for object-level attributes, relations and object-set-level complex concepts provides not only means for generalization of high-level reasoning abilities, but also supports a new form of explainability, as expressed by the DSL and the generated program. Finally, it is essential to measure the quality of explanations, e.g. with the Systems Causability Scale [13], where we need the notion of Causability [15]. In the same way that usability measures the quality of use, causability measures the quality of explanations. This will be urgently needed if explainable AI will not remain a pure theoretical field but a practical relevant field for industry and society.

Acknowledgements. The authors declare that there are no conflict of interests and the work does not raise any ethical issues. Parts of this work has been funded by the Austrian Science Fund (FWF), Project: P-32554 "A reference model of explainable Artificial Intelligence for the Medical Domain", and the Parts of this work have been funded by the European Union's Horizon 2020 research and innovation program under grant agreement No 826078 "Feature Cloud".

References

1. Andreas, J., Rohrbach, M., Darrell, T., Klein, D.: Learning to compose neural networks for question answering. arXiv preprint arXiv:1601.01705 (2016)
2. Anthony, T., Tian, Z., Barber, D.: Thinking fast and slow with deep learning and tree search. In: Guyon, I., Luxburg, U.V., Bengio, S., Wallach, H., Fergus, R., Vishwanathan, S., Garnett, R. (eds.) Advances in Neural Information Processing Systems, pp. 5360–5370. NIPS Foundation (2017)
3. Bennetot, A., Laurent, J.L., Chatila, R., Díaz-Rodríguez, N.: Towards explainable neural-symbolic visual reasoning. In: NeSy Workshop IJCAI (2019)

4. Besold, T.R., Garcez, A.., Bader, S., Bowman, H., Domingos, P., Hitzler, P., Kühnberger, K.U., Lamb, L.C., Lowd, D., Lima, P.M.V., et al.: Neural-symbolic learning and reasoning: A survey and interpretation. arXiv preprint arXiv:1711.03902 (2017)
5. Chen, X., Fang, H., Lin, T.Y., Vedantam, R., Gupta, S., Dollár, P., Zitnick, C.L.: Microsoft coco captions: Data collection and evaluation server. arXiv preprint arXiv:1504.00325 (2015)
6. Dong, H., Mao, J., Lin, T., Wang, C., Li, L., Zhou, D.: Neural logic machines. arXiv preprint arXiv:1904.11694 (2019)
7. Gupta, N., Lin, K., Roth, D., Singh, S., Gardner, M.: Neural module networks for reasoning over text. arXiv preprint arXiv:1912.04971 (2019)
8. Hägele, M., Seegerer, P., Lapuschkin, S., Bockmayr, M., Samek, W., Klauschen, F., Binder, A.: Resolving challenges in deep learning-based analyses of histopathological images using explanation methods (2019)
9. Han, C., Mao, J., Gan, C., Tenenbaum, J., Wu, J.: Visual concept-metaconcept learning. In: Advances in Neural Information Processing Systems, pp. 5002–5013 (2019)
10. He, K., Gkioxari, G., Dollár, P., Girshick, R.: Mask R-CNN. In: Proceedings of the IEEE International Conference on Computer Vision, pp. 2961–2969 (2017)
11. He, K., Zhang, X., Ren, S., Sun, J.: Deep residual learning for image recognition. In: Proceedings of the IEEE Conference on Computer Vision and Pattern Recognition, pp. 770–778 (2016)
12. Hendricks, L.A., Akata, Z., Rohrbach, M., Donahue, J., Schiele, B., Darrell, T.: Generating visual explanations. In: European Conference on Computer Vision, pp. 3–19. Springer (2016)
13. Holzinger, A., Carrington, A., Müller, H.: Measuring the quality of explanations: the system causability scale (SCS). comparing human and machine explanations. KI - Künstliche Intelligenz (German Journal of Artificial intelligence) (2020, in print). https://arxiv.org/abs/1912.09024. Special Issue on Interactive Machine Learning, Edited by Kristian Kersting, TU Darmstadt **34**(2)
14. Holzinger, A., Kickmeier-Rust, M., Müller, H.: Kandinsky patterns as IQ-test for machine learning. In: International Cross-Domain Conference for Machine Learning and Knowledge Extraction, pp. 1–14. Springer (2019)
15. Holzinger, A., Langs, G., Denk, H., Zatloukal, K., Müller, H.: Causability and explainability of AI in medicine. Wiley Interdisciplinary Reviews: Data Mining and Knowledge Discovery (2019). https://doi.org/10.1002/widm.1312
16. Hudson, D., Manning, C.D.: Learning by abstraction: the neural state machine. In: Advances in Neural Information Processing Systems, pp. 5901–5914 (2019)
17. Johnson, J., Gupta, A., Fei-Fei, L.: Image generation from scene graphs. In: Proceedings of the IEEE Conference on Computer Vision and Pattern Recognition, pp. 1219–1228 (2018)
18. Johnson, J., Hariharan, B., van der Maaten, L., Fei-Fei, L., Lawrence Zitnick, C., Girshick, R.: Clevr: a diagnostic dataset for compositional language and aelementary visual reasoning. In: Proceedings of the IEEE Conference on Computer Vision and Pattern Recognition, pp. 2901–2910 (2017)
19. Johnson, J., Hariharan, B., van der Maaten, L., Hoffman, J., Fei-Fei, L., Lawrence Zitnick, C., Girshick, R.: Inferring and executing programs for visual reasoning. In: Proceedings of the IEEE International Conference on Computer Vision, pp. 2989–2998 (2017)

20. Johnson, J., Krishna, R., Stark, M., Li, L.J., Shamma, D., Bernstein, M., Fei-Fei, L.: Image retrieval using scene graphs. In: Proceedings of the IEEE Conference on Computer Vision and Pattern Recognition, pp. 3668–3678 (2015)
21. Kahneman, D.: Thinking, Fast and Slow. Macmillan, New York (2011)
22. Karpathy, A., Fei-Fei, L.: Deep visual-semantic alignments for generating image descriptions. In: Proceedings of the IEEE Conference on Computer Vision and Pattern Recognition, pp. 3128–3137 (2015)
23. Kuhnle, A., Copestake, A.: Shapeworld-a new test methodology for multimodal language understanding. arXiv preprint arXiv:1704.04517 (2017)
24. Lai, F., Xie, N., Doran, D., Kadav, A.: Contextual grounding of natural language entities in images. arXiv preprint arXiv:1911.02133 (2019)
25. Lin, T.Y., Maire, M., Belongie, S., Hays, J., Perona, P., Ramanan, D., Dollár, P., Zitnick, C.L.: Microsoft coco: Common objects in context. In: European Conference on Computer Vision, pp. 740–755. Springer (2014)
26. Liu, Y., Wu, Z., Ritchie, D., Freeman, W.T., Tenenbaum, J.B., Wu, J.: Learning to describe scenes with programs (2018)
27. Manhaeve, R., Dumancic, S., Kimmig, A., Demeester, T., De Raedt, L.: Deep-problog: eural probabilistic logic programming. In: Advances in Neural Information Processing Systems, pp. 3749–3759 (2018)
28. Mao, J., Gan, C., Kohli, P., Tenenbaum, J.B., Wu, J.: The neuro-symbolic concept learner: Interpreting scenes, words, and sentences from natural supervision. arXiv preprint arXiv:1904.12584 (2019)
29. Marcus, G.: The next decade in ai: Four steps towards robust artificial intelligence. arXiv preprint arXiv:2002.06177 (2020)
30. Merdivan, E., Singh, D., Hanke, S., Holzinger, A.: Dialogue systems for intelligent human computer interactions. Electron. Notes Theor. Comput. Sci. **343**, 57–71 (2019). https://doi.org/10.1016/j.entcs.2019.04.010
31. Pohn, B., Mayer, M.C., Reihs, R., Holzinger, A., Zatloukal, K., Müller, H.: Visualization of histopathological decision making using a roadbook metaphor. In: 2019 23rd International Conference Information Visualisation (IV), pp. 392–397. IEEE (2019)
32. Rocktäschel, T., Riedel, S.: Learning knowledge base inference with neural theorem provers. In: Proceedings of the 5th Workshop on Automated Knowledge Base Construction, pp. 45–50 (2016)
33. Santoro, A., Raposo, D., Barrett, D.G., Malinowski, M., Pascanu, R., Battaglia, P., Lillicrap, T.: A simple neural network module for relational reasoning. In: Advances in Neural Information Processing Systems, pp. 4967–4976 (2017)
34. Saranti, A., Taraghi, B., Ebner, M., Holzinger, A.: Insights into learning competence through probabilistic graphical models, pp. 250–271. Springer/Nature, Cham (2019). https://doi.org/10.1007/978-3-030-29726-8_16
35. Torralba, A., Efros, A.A., et al.: Unbiased look at dataset bias. In: CVPR. vol. 1, p. 7. Citeseer (2011)
36. Velik, R., Bruckner, D.: Neuro-symbolic networks: introduction to a new information processing principle. In: 2008 6th IEEE International Conference on Industrial Informatics, pp. 1042–1047. IEEE (2008)
37. Yi, K., Gan, C., Li, Y., Kohli, P., Wu, J., Torralba, A., Tenenbaum, J.B.: Clevrer: Collision events for video representation and reasoning. arXiv preprint arXiv:1910.01442 (2019)
38. Yi, K., Wu, J., Gan, C., Torralba, A., Kohli, P., Tenenbaum, J.: Neural-symbolic VQA: disentangling reasoning from vision and language understanding. In: Advances in Neural Information Processing Systems, pp. 1031–1042 (2018)

The Impact of Supercategory Inclusion on Semantic Classifier Performance

Piotr Borkowski(✉), Krzysztof Ciesielski, and Mieczysław A. Kłopotek

Institute of Computer Science, Polish Academy of Sciences,
ul. Jana Kazimierza 5, Warszawa, Poland
{piotrb,kciesiel,klopotek}@ipipan.waw.pl
https://ipipan.waw.pl/

Abstract. It is a known phenomenon that text document classifiers may benefit from inclusion of hypernyms of the terms in the document. However, this inclusion may be a mixed blessing because it may fuzzify the boundaries between document classes [5,6,10].

We have elaborated a new type of document classifiers, so called semantic classifiers, trained not on the original data but rather on the categories assigned to the document by our semantic categorizer [1,4], that require significantly smaller corpus of training data and outperforms traditional classifiers used in the domain.

With this research we want to clarify what is the advantage/disadvantage of using supercategories of the assigned categories (an analogon of hypernyms) on the quality of classification. In particular we concluded that supercategories should be added with restricted weight, for otherwise they may deteriorate the classification performance. We found also that our technique of aggregating the categories counteracts the fuzzifying of class boundaries.

Keywords: Text mining · Semantic gap · Semantic similarity · Document categorization · Document classification · Category aggregation · Supercategory inclusion

1 Introduction

The text document classification is used as a supporting tool in a number of areas in business and administration. Let just mention classification of customer feedback, of customer questions (for forwarding them to experts), of offer document content, of technical requirements to different engineering area, of emails into spam/non-spam, of books (libraries, shops) based on their abstracts into librarian categories, of applicant CVs, query classification and clustering etc.

The respective methods are based usually on data mining techniques (like Naive Bayes, Wide-Margin Winnow, L-LDA, etc.) that are able to handle long input data records. Though various methods proved useful both in data mining and in text mining, there occurs one important drawback for text mining: the meaning and the value range of individual attributes of an object are not well

M. Stettinger et al. (Eds.): ISMIS 2020, SCI 949, pp. 69–79, 2021.
https://doi.org/10.1007/978-3-030-67148-8_6

defined in text mining, because content may be expressed in different ways, using different words, while the same word can express different things. This fact impacts heavily the amount of labeled data needed for classifier training. The problem becomes more grievant if there occurs the problem of so-called semantic gap. The semantic gap means roughly that the vocabulary of the training set and the test set differs syntactically, though is similar semantically. In such cases the traditional classification methods would fail nearly by definition. One can easily guess that understanding the semantics of documents would be helpful or even indispensable.

We applied this heuristics when developing the document classification method SemCla (*Semantic Classifier*) based on the semantic categorizer SemCat, see [1,4]. The idea of the approach is to characterize a document by a set of categories (from Wikipedia ($\mathfrak{W}$) category hierarchy) instead of original bag of words, and then to classify new objects based on similarity to labeled objects. This approach has the basic advantage that we go beyond the actual formulation of the document text and use rather its semantics, the conceptual representation. The disadvantage is of course the risk that the generalization of a document to its categorical description may prove too broad depending on the generality of the description applied.

In this research we investigate the impact of the generality of categories by usage of which the documents are described onto the accuracy of classification.

The outline of the paper is as follows: in Sect. 2 we recall previous works on closing a semantic gap. In Sect. 3 we describe our semantic classifier SemCla. Section 4 is devoted to the idea of the semantic categorizer SemCat. Section 5 presents the measures of semantic similarity used by the SemCla. We describe the experimental setup in Sect. 7. Then in Sect. 8 we present and discuss the results of an empirical investigation. Section 9 contains conclusions from our work and envisaged possibilities of further research.

2 Previous Work on Closing Semantic Gap

The issue of "semantic gap" has been investigated in the past by a number of researchers. There exist subtle differences in understanding the problem, nonetheless there is a general consensus on its severity. We focus on the aspect encountered in text retrieval where data come for different domains.

A detailed overview of *cross-domain text categorization* problem was presented in the paper [9]. It seems to be a very common case in practical tasks that the training and the test data originate from different distributions or domains. Many algorithms have been developed or adapted for such a setting. Let us just mention such conventional algorithms like: Rocchio's Algorithm, Decision Trees like: CART, ID3, C4.5; Naive Bayes classifier, KNN, Support Vector Machines (SVM). But there exist also novel cross-domain classification algorithms: Expectation-Maximization Algorithm, CFC Algorithm, Probabilistic Latent Semantic Analysis (PLSA), Latent Dirichlet Allocation (LDA), Co-cluster based Classification Algorithm [11]. In [3] PLSA related ideas are used for document modeling via combining semantic resources and statistically extracted topics.

The paper [7] focuses on a general overview of semantic gap issues in information retrieval. Authors discuss among other text mining and retrieval. They study reorganizing search results by applying a post-retrieval clustering system. They enhance search results ("snippets") by adding so called *topics*. A topic is understood as a set of words produced by PLSA or LDA on some external data collection. They cluster or label the snippets enriched with topics.

Authors of [8] improve categorization by adding semantic knowledge from Wikitology (knowledge repository based on $\mathfrak{W}$). They applied diverse text representation and enrichment methods and used SVM to train a classification model.

Another approach to categorization, on which we base our research, is described in [1,4]. We recall it in more detail in the next sections.

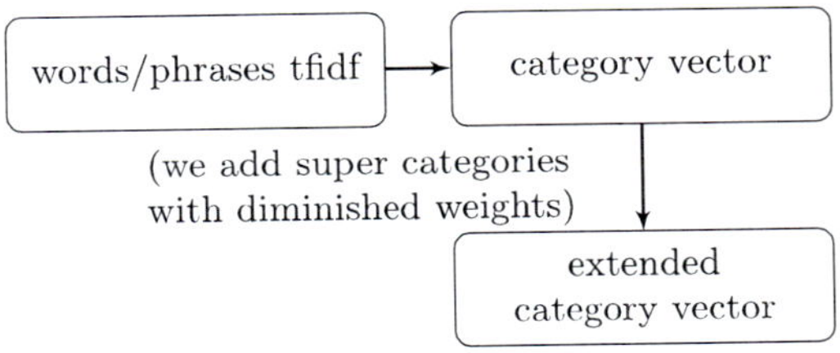

Fig. 1. Single document category representation

3 Semantic Classification Method SemCla

We will briefly present the new *semantic classifier* SemCla, introduced in [1]. It is based on a category representation of a document produced by SemCat (see Sect. 4.1), used in combination with semantic measures discussed in Sect. 5.

3.1 Outline of the Algorithm

SemCat derives a list of categories with weights, based on the words and phrases from the document. This new document representation can be viewed as a vector of weights for all $\mathfrak{W}$ categories. Hence it shall be called *vector of categories*. We use it to calculate cosine product as a measure of document similarity. It turned out that the algorithm performs better when for each category from the vector of categories a super category (from $\mathfrak{W}$ hierarchy) is added with weight equal the initial weight multiplied by a constant α. Thus we obtain the *extended category vector*. This process is visualized in Fig. 1. The semantic classification is performed in the way described below and illustrated in Fig. 2.

1. Categorization of documents from training and test sets via SemCat to obtain category vectors that represent their content.
2. Extension of the category vectors for all documents by adding a super category (according to $\mathfrak{W}$ hierarchy) with weight equal the initial weight multiplied by the constant α (extended category vectors are created).
3. Optionally creation of a centroid for each document groups from training set (the average of category weight vectors of group elements, normalized to unit length).
4. Classification of a new document (represented by its extended category vector) by finding the nearest group (in the sense of the cosine product) in the training set.

The above algorithm is parameterized by three quantities: supercategory importance parameter $\alpha \in [0,1]$, the switch *adapagg* $\in \{True, False\}$ telling whether to use the automated category aggregation within SemCat and the method of identification of the nearest group $neargr \in \{All, Centroid\}$. Option $neargr = All$ means that the group is chosen where the average (cosine) similarity to each group element is the smallest, while $neargr = Centroid$ means that the similarity to the group centroid is taken into account only. The latter method of classification of a new element is faster, while a bit less precise. It is known from earlier research on enriching document representation with hypernyms that there exist some danger of fuzzifying the document class boundaries. Therefore, in this study, we investigate the impact of various choices of the α coefficient and the *adapagg* switch on the accuracy of classification.

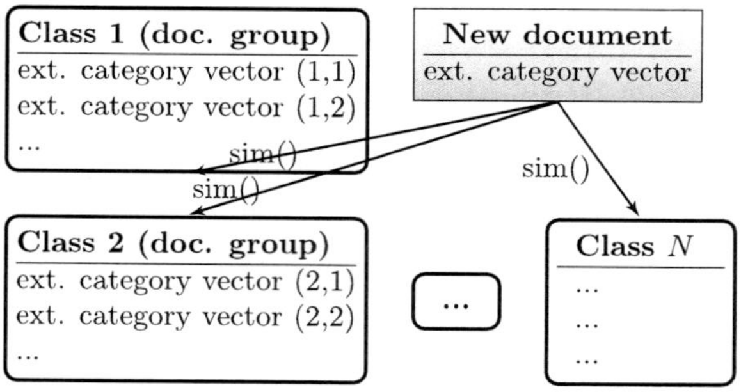

Fig. 2. Categorization as a classification (SemCla algorithm)

4 Semantic Categorization Method SemCat

The taxonomy-based categorization method SemCat was described in detail in [1,4]. Below we present only its brief summary.

4.1 Outline of the Algorithm

The algorithm exploits a taxonomy of categories (a directed acyclic graph with one root category) like Wikipedia ($\mathfrak{W}$) category graph or *Medical Subject Headings* (MeSH) ontology[1], the goal of which is to provide with semantic (domain) information. The taxonomy must be connected to a set of concepts. It is assumed that a document is a "bag of concepts". Every concept needs to be linked to one or more categories. Every category and concept is tagged with a string label. Strings connected with categories are used as an outcome for the user. And those attached to concepts are used for mapping a text of document into the set of concepts.

For the experimental design we used $\mathfrak{W}$ category graph with the concept set of $\mathfrak{W}$ pages. Tags for $\mathfrak{W}$ categories were their original string names. Set of string tags connected with a single $\mathfrak{W}$ page consists of: lemmatized page name and all names of disambiguation pages that link to that page. Categorization of a document encompasses the following steps: removal of stop words and very rare/frequent words, lemmatizing, finding phrases and calculating normalized *tfidf* weights for terms and phrases. Calculation of a standard *term frequency inverse document frequency* is based on word frequencies from the collection of all $\mathfrak{W}$ pages. The next step is to map document terms and phrases into a set of concepts. In the case of homonyms, disambiguation procedure is applied to the concept assignment: we select the concept that is the nearest by similarity measure defined by Eqs. (1) and (2) (see Sect. 5) to the set of concepts that was mapped in an unambiguous way.

When every term in the document is assigned to a proper concept ($\mathfrak{W}$ page), then all concepts are mapped to $\mathfrak{W}$ categories. In this way usually one term maps to more than one category. The weight associated to that term transferred proportionally to all its categories. Sum of weights assigned to the categories equals to the sum of *tfidf* for terms. The outcome of that procedure is a ranked list of categories with weights. In the last step either automated aggregation is applied to the weighted ranking ($adapagg = True$, see Sect. 6) and/or top-N categories are chosen out of it.

5 Similarity Measures

The semantic similarity measures used above are based on: the unary function *IC* (*Information Content*) and binary function *MSCA* (*Most Specific Common Abstraction*). Their inputs are categories from a taxonomy. The measures have been introduced in [4]. Let us only recall the formulas. For a category k let $IC(k) = 1 - log\,(1 + s_k)\,/log\,(1 + N)$, where s_k is the number of taxonomy concepts in the category k and all its subcategories, and N is the total number of taxonomy concepts. For two given categories k_1 and k_2 let $MSCAIC(k_1, k_2) = \max\{IC(k) : k \in CA(k_1, k_2)\}$ where $CA(k_1, k_2)$ is the set of super-categories

[1] https://www.nlm.nih.gov/mesh/.

for both categories k_1 and k_2. Then define $MSCA(k_1, k_2) = \{k : IC(k) = MSCAIC(k_1, k_2)\}$. Define Lin and Pirro-Seco similarity:

$$sim_{\mathrm{Lin}}(k_1, k_2) = \frac{2 \cdot MSCAIC(k_1, k_2)}{IC(k_1) + IC(k_2)} \tag{1}$$

$$sim_{\mathrm{PirroSeco}}(k_1, k_2) = \frac{1}{3}\Big(3 \cdot MSCAIC(k_1, k_2) - IC(k_1) - IC(k_2) + 2\Big) \tag{2}$$

Similarity between pages p_i and p_j is computed by aggregation of similarity between each pair of categories (k_i, k_j) such that p_i belongs to the category k_i and p_j to k_j:

$$sim_{\mathrm{PAGE}}(p_i, p_j) = max\{sim_{\mathrm{CAT}}(k_i, k_j) : p_i \in k_i \wedge p_j \in k_j\} \tag{3}$$

6 Unsupervised Adaptive Aggregation of Categories

The advantage of using categories instead of words from the document is the bridging of the semantic gap. However, due to ambiguity of the document words, there is a risk of adding categories that may not be related to the main topic of the document, and thus fuzzifying its content which harms the effort classification. Focusing on the main topic of the document of the collection may be helpful in such a case, as discussed already in [2]. The topic description may be provided either manually or it has to be deduced from document content. The algorithm developed in [2] handles both cases (see below).

6.1 Mapping to the Predefined Set of Labels

Consider an aggregation algorithm generalizing the original ranking of categories $k_1, k_2, \ldots, k_R$ by transforming it to the set of **manually** selected target labels $L = \{l_1, l_2, \ldots, l_T\}$. The purpose of the algorithm is to assign a weight to each of the target labels so that the total weight of the original and target categories remains the same (i.e. original weights are redistributed).

Let the category k_i with original weight w_i be a sub-category (not necessarily direct) of a subset of the target categories, $l_{k_i}^1, l_{k_i}^2, \ldots, l_{k_i}^S$. Then each of the target categories has its weight increased by w_i/S.

This procedure is applied to each original category k_i, $i = 1, \ldots, R$, and the propagated weights are summed up at the target categories, inducing their ranking.

6.2 Unsupervised Mapping

If not available, the target set $l_1, l_2, \ldots, l_T$ may be constructed in unsupervised manner, based on the original categories and their weights (k_i, w_i) $i = 1, \ldots, R$. Only the parameter T is supplied by a user. For a given set of input categories $K_R = \{k_1, k_2, \ldots, k_R\}$ construct a set of all its $MSCA$ categories and denote it by $M(K_R) = \{MSCA(k_i, k_j);\ 1 \le i < j \le T\}$. Apply this recurrently: $M_2(K_R) = M(M(K_R))$, etc. until a singleton set (M_S) is obtained, $M_S = M(\cdots(M(K_R)))$. All these sets are added, providing candidate superset of target categories: $\mathscr{M} = M(K_R) \cup M_2(K_R) \cup \ldots \cup M_S$. For a given T choose a subset $\{l_1, l_2, \ldots, l_T\} \subset \mathscr{M}$ as follows:

1. For each category $l \in \mathscr{M}$ compute $weight(l) = \sum_{i=1}^{R} w_i \cdot sim(k_i, l)$, where sim is defined either by Eq. (2) or (1).
2. Sort all the categories according to the descending value of $weight(l)$.
3. Choose the first T categories, obtaining subset $\mathscr{M}' \subset \mathscr{M}$ as the target set of categories.

We set then $L := \mathscr{M}'$ and proceed as in Sect. 6.1.

All the presented algorithms are of linear complexity in the number of the documents (and their length). The complexity of aggregation process is limited by the number of categories in the original ranking and by the length of the longest path in the Wikipedia category graph.

7 Experimental Setup

We performed three types of experiments: investigation of the impact of (1) the supercategory importance parameter α, (2) the aggregation, and (3) the choice of centroids as group representative on the performance of a semantic classifier SemCla.

The benchmark was made of documents downloaded from various news pages. The training and evaluation parts were taken from separate collections to achieve different wordings in each of them. The training set consists of news from the popular science portal kopalniawiedzy.pl merged with documents from one directory from forsal.pl (the domain about finance and economy). In this way, a collection of documents belonging to 7 topical classes was created. The training set has the following characteristics (classes are indicated in bold):

- documents from kopalniawiedzy.pl: **astronomy-physics** $N = 311$; **humanities** $N = 244$; **life science** $N = 3222$; **medicine** $N = 3037$; **psychology** $N = 1758$; **technology** $N = 6145$;
- documents from forsal.pl from the directory *Giełda* (Stock exchange): **business** $N = 1986$.

For evaluation we downloaded directories from www.rynekzdrowia.pl (containing medical news – labeled as **medicine**) and merged it with economical documents from www.forsal.pl and www.bankier.pl (market, finances, business – labeled as **business**). The following datasets were used for evaluation:

- directories from www.rynekzdrowia.pl: Ginekologia (Gynecology): **medicine** $N = 1034$; Kardiologia (Cardiology): **medicine** $N = 239$; Onkologia (Oncology): **medicine** $N = 1195$,
- directories from www.forsal.pl: Waluty (Currencies): **business** $N = 2161$; Finanse (Finances): **business** $N = 1991$,
- documents from www.bankier.pl: **business** $N = 978$.

7.1 Efficiency Measures

To assess the impact of algorithm parameters, we used standard *accuracy* measure: $acc_{0\text{-}1}(x, y) = \mathbf{1}_x(y)$.

8 Results

The results of the experiments have been summarized in Tables 1, 2, 3 and 4. The Table 1 presents classification experiments where similarity to all group elements was considered and no category aggregation was performed. The Table 2 differs from previous experimental setup in that automated category aggregation described in Section 6.2 was applied. The Tables 3 and 4 show the case of application of similarity of new elements to the group centroids instead of all group elements (see Sect. 3.1) in cases described by Tables 1 and 2. The columns of each table represent the impact of weight of the supercategories (the parameter α). The rows of the tables indicate the accuracy for individual classes of documents. The Table 1 contains additionally (two last columns) results of application of standard classifiers Bayes and Wide-Margin Winnow using the terms from the documents for training and classification.

It turns out that inclusion of supercategories improves generally the classification accuracy, but only for smaller impact factor α of the supercategory. α values between $\frac{1}{3}$ and $\frac{1}{2}$ appear to be optimal across all topical classes. If the factor α approaches 1, the performance is even worse (in most cases) than in case without considering supercategories ($\alpha = 0$). Note also that for SemCat with centroid and without aggregation, the adding of supercategories deteriorates the results (Table 3). Category aggregation appears to be a mixed blessing. In case of no supercategory ($\alpha = 0$) generally worse results are achieved. However, with increase of α the effects of category aggregation seem to be positive. But note that the computation was performed for the parameter $T = 3$ which means that the whole text describing a document was "compressed" into three categories only and the classification results are worse only by 5–20% points, and via extension by supercategories, the accuracy is restored or even beaten compared to the case when aggregation is not performed. As could be expected, the replacement of similarity to all with the similarity to the centroids deteriorates the classification accuracy, though the advantage is that classification of a new element is sped up by a factor equal to the average class cardinality. Adding supercategories does not compensate in general this switch between modes of group identification. Nonetheless, adding supercategories improves in most cases the accuracy in this version of algorithm also. Let us briefly mention the comparison to two broadly known text classification methods: (Naive) Bayes and Winnow which are based on the document terms. They perform worse in all classes than the best performing version of SemCla in each class for all based group identification (Tables 1, 2), but are superior in all cases to centroid-based SemCla without category aggregation (Table 3). Category aggregation improves the situation to the extent that Winnow is defeated by SemCla (Table 4).

Table 1. The average values of accuracy measure for data with a semantic gap. Columns 1-7 are the results for SemCla with various values of α parameter. Last two columns: results for the Bayes and Wide-Margin Winnow classifiers.

	α								
	0.0	0.16	0.33	0.5	0.66	0.83	1.0	Bayes	W-M Winnow
Bankier (Business)	0.741	0.791	**0.841**	0.820	0.712	0.563	0.446	0.701	0.611
Forsal (Currencies)	0.984	0.990	**0.994**	**0.994**	0.990	0.970	0.887	0.987	0.924
Forsal (Finances)	0.966	0.981	**0.983**	0.981	0.975	0.951	0.872	0.959	0.925
Gynecology	0.612	0.702	**0.801**	0.794	0.662	0.476	0.358	0.617	0.581
Cardiology	0.864	0.902	**0.942**	0.931	0.901	0.854	0.752	0.916	0.891
Oncology	0.816	0.858	**0.883**	0.863	0.803	0.671	0.528	0.84	0.856

Table 2. The average values of accuracy measure for data with a semantic gap for SemCla using **automated category aggregation** ($T = 3$) with various values of α parameter.

	$\alpha = 0.0$	$\alpha = 0.16$	$\alpha = 0.33$	$\alpha = 0.5$	$\alpha = 0.66$	$\alpha = 0.83$	$\alpha = 1.0$
Bankier (Business)	0.626	0.757	0.813	**0.852**	0.832	0.732	0.568
Forsal (Currencies)	0.848	0.985	0.994	**0.995**	0.993	0.987	0.957
Forsal (Finances)	0.935	0.962	0.971	**0.978**	0.977	0.970	0.928
Gynecology	0.456	0.853	**0.858**	0.832	0.743	0.537	0.355
Cardiology	0.481	0.944	**0.966**	0.950	0.917	0.867	0.753
Oncology	0.723	0.923	**0.928**	0.906	0.851	0.737	0.546

Table 3. The average values of accuracy measure for data with a semantic gap for SemCla using **centroid** with various values of α parameter.

	$\alpha = 0.0$	$\alpha = 0.16$	$\alpha = 0.33$	$\alpha = 0.5$	$\alpha = 0.66$	$\alpha = 0.83$	$\alpha = 1.0$
Bankier (Business)	**0.515**	0.511	0.478	0.381	0.319	0.276	0.238
Forsal (Currencies)	**0.859**	0.788	0.745	0.627	0.502	0.366	0.284
Forsal (Finances)	**0.883**	0.801	0.766	0.628	0.501	0.369	0.293
Gynecology	**0.469**	0.446	0.406	0.285	0.217	0.176	0.141
Cardiology	**0.801**	0.684	0.611	0.475	0.354	0.272	0.217
Oncology	**0.775**	0.706	0.626	0.452	0.306	0.213	0.169

Table 4. The average values of accuracy measure for data with a semantic gap for SemCla using **automated category aggregation** ($T = 3$) and **centroid** with various values of α parameter.

	$\alpha = 0.0$	$\alpha = 0.16$	$\alpha = 0.33$	$\alpha = 0.5$	$\alpha = 0.66$	$\alpha = 0.83$	$\alpha = 1.0$
Bankier (Business)	0.558	**0.663**	0.662	0.627	0.570	0.503	0.426
Forsal (Currencies)	0.783	0.971	**0.980**	0.978	0.965	0.928	0.846
Forsal (Finances)	0.907	0.946	**0.952**	0.949	0.932	0.892	0.812
Gynecology	0.335	**0.593**	0.562	0.474	0.350	0.245	0.149
Cardiology	0.355	0.791	**0.832**	0.795	0.737	0.642	0.474
Oncology	0.616	**0.798**	0.794	0.734	0.611	0.468	0.350

9 Conclusions

In this paper we investigated the impact of inclusion of supercategories when classifying documents based on their semantic categories. Usage of semantic categories is know to be superior to the usage of words/terms from the document in case of the so-called semantic gap. However, it has not been studied whether or not the inclusion of supercategories may be beneficial in such applications. It was worth investigating because it is already know that text document classifiers may benefit from inclusion of hypernyms of the terms in the document, though in some cases it may fuzzify the boundaries between document classes and hence the effect may be contrary to the desired one. Our investigation shows that in fact we need to weigh carefully the importance of the supercategories in order to gain from their usage. By adding them with a weight of $\frac{1}{3}$ of the original category weight one usually benefits most.

Note also that the very idea of replacing the document text with the corresponding categories introduces in fact "superterms" of the terms of the original document. While their advantage for handling semantic gap is obvious, one can ask whether or not they introduce also too much noise like the too broad hypernyms. This was investigating by exploiting the previously developed method of automated category aggregation. An extreme approach has been applied where the number of categories was reduced to 3 (a very significant document compression). This compression lead to worsening of classification accuracy unless supercategories are included. This may constitute a hint that category aggregation methods should take into account supercategory inclusion. Nonetheless even without supercategories, it is worth mentioning that the extreme reduction of the document description did't deteriorate the classification accuracy much.

It is an open question, what would be the optimal number T of categories that should be used in the document compression, and how it depends to the number of classes into which the classification is to be performed.

In this paper we have considered a very laborious and a very simple way of deciding group membership for new objects. While the laborious method seems to be quite accurate, the simple one is not. It is therefore an open issue how to modify the latter in order to be still efficient but without deviating too much from the accuracy of the laborious method.

This research opens up a number of further interesting areas of research. Semantic approach (in its base, unsupervised setting) could be tested also for clustering tasks under semantic gap scenario as well as to mixtures of classification and clustering.

References

1. Borkowski, P.: Metody semantycznej kategoryzacji w zadaniach analizy dokumentów tekstowych. Ph.D. thesis, Institute of Computer Science of Polish Academy of Sciences (2019)
2. Borkowski, P., Ciesielski, K., Kłopotek, M.A.: Unsupervised aggregation of categories for document labelling. In: Foundations of Intelligent Systems - 21st International Symposium. ISMIS 2014, Roskilde, Denmark, 25–27 June 2014. Proceedings, pp. 335–344 (2014)
3. Chemudugunta, C., Holloway, A., Smyth, P., Steyvers, M.: Modeling documents by combining semantic concepts with unsupervised statistical learning. In: Sheth, A., et al. (eds.) The Semantic Web - ISWC 2008. LNCS, vol. 5318, pp. 229–244. Springer, Berlin (2008)
4. Ciesielski, K., Borkowski, P., Kłopotek, M.A., Trojanowski, K., Wysocki, K.: Wikipedia-based document categorization. In: SIIS 2011, pp. 265–278 (2011)
5. Huang, Z., Thint, M., Qin, Z.: Question classification using head words and their hypernyms. In: EMNLP 2008: Proceedings of the Conference on Empirical Methods in Natural Language, pp. 927–936, October 2008
6. Li, X., Roth, D.: Learning question classifiers. In: The 19th International Conference on Computational Linguistics, vol. 1, pp. 1–7 (2002)
7. Nguyen, C.T.: Bridging semantic gaps in information retrieval: context-based approaches. ACM VLDB **10** (2010)
8. Rafi, M., Hassan, S., Shaikh, M.S.: Content-based text categorization using wikitology. CoRR **abs/1208.3623** (2012)
9. Ramakrishna Murty, M., Murthy, J., Prasad Reddy, P., Satapathy, S.: A survey of cross-domain text categorization techniques. In: RAIT 2012, pp. 499–504. IEEE (2012)
10. Scott, S., Matwin, S.: Text classification using wordnet hypernyms. In: Use of WordNet in Natural Language Processing Systems: Proceedings of the Conference, pp. 38–44, 45–52. Association for Computational Linguistics (1998)
11. Wang, P., Domeniconi, C., Hu, J.: Using Wikipedia for co-clustering based cross-domain text classification. In: ICDM 2008, pp. 1085–1090. IEEE (2008)

Recognition of the Flue Pipe Type Using Deep Learning

Damian Węgrzyn[1(✉)], Piotr Wrzeciono[2], and Alicja Wieczorkowska[1]

[1] Polish-Japanese Academy of Information Technology, Koszykowa 86, Warsaw, Poland
damian@wegrzyn.info, alicja@poljap.edu.pl
[2] Warsaw University of Life Sciences, Nowoursynowska 166, Warsaw, Poland
piotr_wrzeciono@sggw.edu.pl

Abstract. This paper presents the usage of deep learning in flue pipe type recognition. The main thesis is the possibility of recognizing the type of labium based on the sound generated by the flue pipe. For the purpose of our work, we prepared a large data set of high-quality recordings, carried out in an organbuilder's workshop. Very high accuracy has been achieved in our experiments on these data using Artificial Neural Networks (ANN), trained to recognize the details of the pipe mouth construction. The organbuilders claim that they can distinguish the pipe mouth type only by hearing it, and this is why we decided to verify if it is possible to train ANN to recognize the details of the organ pipe, as this confirms a possibility that a human sense of hearing may be trained as well. In the future, the usage of deep learning in the recognition of pipe sound parameters may be used in the voicing of the pipe organ and the selection of appropriate parameters of pipes to obtain the desired timbre.

Keywords: Flue pipe · Deep learning · Labium recognition

1 Introduction

A pipe organ consists of many pipes of various types, collected in ranks and stops. The majority of pipes is divided into two groups according to the method of sound generation: flue (labial) pipes and reed pipes [1]. Flue pipes have a variety of timbres which are achieved by the modifications of the pipe mouth. It is critical to the resulting sound because it influences the pipes' voice which has to be tuned [13]. The procedure of achieving a harmonious overall sound for the whole rank is called voicing [14].

The pipe mouth has a significant impact on a vibrating air jet [2,6,7,16,17]. In the case of flue pipes, there are four additions that can be applied to change the mouth of the pipe: ears, beard, plate, and roller (Fig. 1). These additions do

Partially supported by research funds sponsored by the Ministry of Science and Higher Education in Poland.

M. Stettinger et al. (Eds.): ISMIS 2020, SCI 949, pp. 80–93, 2021.
https://doi.org/10.1007/978-3-030-67148-8_7

not occur together. They make the pipe sound lower and darker. Moreover, the sound generation is smoother and faster. Research confirms that the generated sound has a stronger, more pronounced fundamental frequency, and the amount of the harmonics is increased by these four additions. Another effect is the growth of the level of the fundamental frequency accompanied by the lowered level of other harmonics, and a slight decrease in sound pitch. The organbuilders also mention a perceptible impact of the mouth end correction on the airstream [14].

Fig. 1. Different types of elements added to flue organ pipes (from left): ears, beard, plate, roller.

Experienced organbuilders are able to recognize various stops. The main aim of this paper is to contribute to the verification of the thesis that the trained listener, e.g. organist or organbuilder, can distinguish between types of pipe mouth by listening to the generated steady-state sound. Verifying this thesis using listening tests is very difficult because it requires carrying out tests on a large number of specialists, so such tests are not feasible. Therefore, we used deep learning methods that allow creating ANNs whose structure and functions resemble the work of the human brain. Deep Learning is used in machine learning to perform natural tasks of the human brain, e.g. sound recognition.

2 Methodology

For the purpose of our research, we recorded sounds of organ pipes of various fundamental frequencies in the organbuilder's workshop. We measured the sound level for all pipes recorded. Next, the audio data were analyzed, and used for ANN training, aiming at the recognition of the details of the pipe mouth.

The ANNs have been used in research related to the area of music for over 20 years [4,8,18]. Various types of networks are used in current research, especially the Convolutional Neural Network (CNN) [9] or the Long Short-Term Memory (LSTM) [12]. Thanks to them, it is possible to achieve good results and perform complex music processing tasks.

2.1 Recordings and Measurements

We performed the recordings and measurements in the organbuilder's workshop, on the voicing chest. The air temperature was 18.5 °C. The air pressure in the windchest was set to 80 mm water gauge. The atmospheric pressure in the workshop was 1004 hPa.

The sound measurements were performed before the recording. Each recorded sound was generated by a pipe for about 4 s. Several sounds were recorded for each pipe. The measurements were made for various flue pipes tuned to miscellaneous frequencies, as shown in Table 1.

Table 1. Pipes used in our research.

Labium	Register	No of recordings	Construction
Ears	Principal 4-foot	70	Open, pewter (75% tin, 25% lead)
Beard	Bourdon 8-foot	61	Open, oakwood
	Bourdon 16-foot	33	Stopped, oakwood
	Dolce Flute 8-foot	53	Open, spruce
	Flute 4-foot	86	Open, spruce
Plate	Gamba 8-foot	57	Open, metal (55% lead, 45% tin)
Roller	Bass Principal 16-foot	185	Open, pine
	Geigen Principal 8-foot	156	Open, pine

Recordings were made using one measuring microphone with omnidirectional characteristics, the sensitivity of 10 mV/Pa and an equivalent noise level of 20 dBA. This microphone was positioned at a distance of 37 mm from the mouth (Fig. 2). The measurement system was calibrated using Class 1 acoustic calibrator (1 kHz, 114 dB).

2.2 Data Sets

The data sets used for training, validation and testing of ANN were prepared as follows. In the first stage, the recordings of 700 sounds were transformed

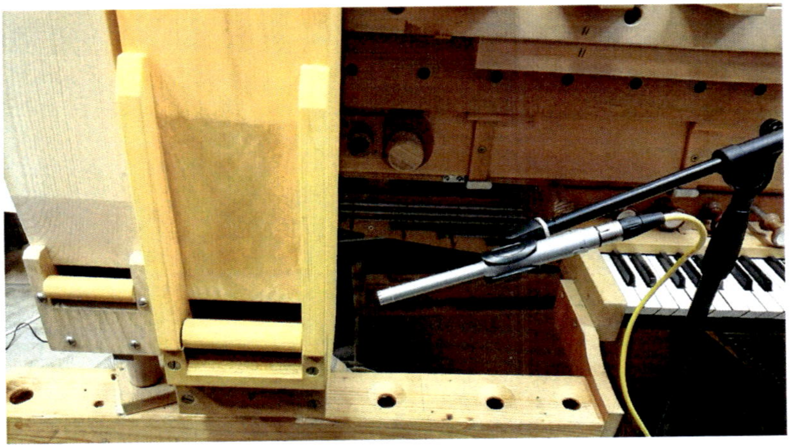

Fig. 2. Microphone setting in the process of pipe sound recording.

into frequency spectra using Fast Fourier Transform (FFT) [11], calculated for a frame length of 2048 samples using the Hamming window (48 kHz sampling rate was applied). One frame from the central part of each sound, representing the steady-state, was selected for further analysis. Examples of power spectra for each flue pipe type (in [dB] scale) are shown in Fig. 3. The obtained frequency components of the spectrum, as well as the level of each component, were saved in a file. The highest frequencies – above 16.5 kHz – have been omitted due to the very low level of these harmonics. This allowed us to reduce the number of inputs, and maintain the stability of training results. For each input record, i.e. for each pipe, one of four output categories has been assigned: ears, beard, plate, or roller.

In the second stage, the collected data representing 700 sounds were randomly divided into three subsets: 400 sounds for training, 150 for validation and 150 for tests. The ANN was trained, validated and tested using these data sets.

2.3 The Structure of the Artificial Neural Network

The ANN model and structure were prepared in the Python programming language using the Scikit-learn library in the Deep Cognition tool [3]. This paper presents the ANN model for which the best accuracy has been achieved.

The data set in the form of 700 records (as described in Sect. 2.2), one record per sound, was used as input. Data were stored in a file, where the 700 columns represent the values of FFT power spectrum for consecutive frequency bins, and the 700 rows represent the analyzed sounds. The output values belong to a set of four categories: ears, beard, plate, roller. They were assigned to each row as a 701st column.

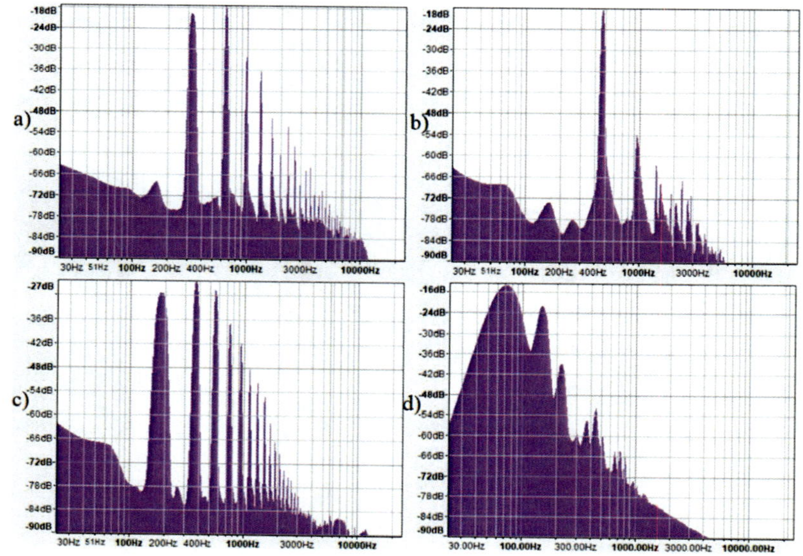

Fig. 3. Spectra of flue pipes: a) ears, b) beard, c) plate, d) roller.

The hidden layer consists of 11 layers, as shown in Table 2. Three types of core layers were used: dense, activation, dropout and one type of normalization layers - batch normalization. For simplicity, both input and output were 700 in each case, except for the last instance, where output is equal to the number of categories.

The dense layer is a regular densely-connected ANN layer with a linear activation function:

$$f(x) = a \times x \tag{1}$$

where: a – the slope of the line.

The activation layer applies an activation function to the output. In all cases, a Rectified Linear Units (ReLU) function was used, which is an approximation of the Softplus function by simple zeroing negative values:

$$f(x) = max(0, x) \tag{2}$$

This procedure speeds up both the implementation and the calculation of the algorithm and significantly accelerates the convergence of the stochastic gradient descent method [10].

The dropout layer randomly bypasses certain neurons during network training. Only part (p) of the layer's neurons are left and the rest are ignored. The method is implemented by applying a binary mask (r_n) to the output values of

Table 2. Scheme of the ANN layers of the best model.

No	Layer	Output dimensions
–	Input	700
1	Dense	700
2	Activation	700
3	Batch normalization	700
4	Dropout	700
5	Dense	700
6	Batch normalization	700
7	Activation	700
8	Dropout	700
9	Dense	700
10	Activation	700
11	Dense	4
–	Output	4

each layer. The r_n mask is different for each layer and is generated with every forward propagation [15]:

$$y_n' = \frac{r_n * y_n}{p} \tag{3}$$

where: y'_n – a modified output vector; y_n – an input vector.

Fraction to drop in dropout layers was set to 0.2.

The batch normalization layer normalizes the activations of the previous layer at each batch. A feature-wise normalization was used during this research, in which each feature map in the input is normalized separately. The batches were set to the size of 32.

The Adadelta optimizer was used during training, which is a per-dimension learning rate method for gradient descent. This method does not require manual tuning of a learning rate and is robust to noisy gradient information, various model architectures, data sets modalities and choices of hyperparameters [19]. The parameters of this optimizer were left at their default values: the initial learning rate was 1 and the rho hyperparameter, which is a decay factor that corresponds to the fraction of gradient to keep at each time step, was set to 0.95.

3 Results

The best ANN model used in our research has achieved high training and validation accuracy. The prepared data set was used in circa 60% for training, 20% for validation and 20% for tests. Weights were saved on each epoch and the validation dataset was used to tune the parameters. The training was limited to 20 epochs because further iterations did not significantly improve accuracy. The best epoch in terms of train accuracy was the last one.

The average train accuracy that was achieved during this research was 0.9116 and the validation average accuracy was 0.9563. The obtained accuracy is shown in Fig. 4.

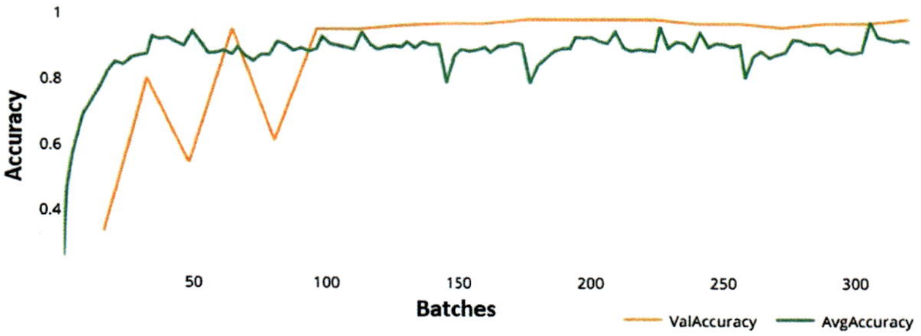

Fig. 4. The best average accuracy (AvgAccuracy) and validation accuracy (ValAccuracy).

During training of the model, the average loss of 1.1433 has been achieved. The validation loss reached near 0.0001. Figure 5 shows the decreasing loss with the increase in the number of batches.

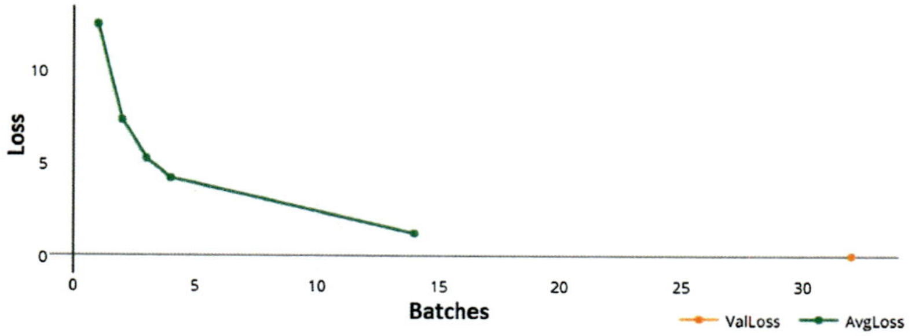

Fig. 5. The best average loss (AvgLoss) and validation loss (ValLoss).

Of the 150 tested flue pipes with various labia, only two were recognized incorrectly, i.e. instead of a beard, the ANN recognized a roller. The confusion matrix presented in Table 3 shows that the selected neural network model obtains very high-quality classification results. The classifier accuracy is 0.987, while the macro average of the F1 score is 0.991.

Table 3. Confusion matrix with classification accuracy, precision, recall and F1 score.

	True ears	True beard	True plate	True roller
Predicted as ears	16	0	0	0
Predicted as beard	0	43	0	0
Predicted as plate	0	0	15	0
Predicted as roller	0	2	0	74
Classification accuracy	**0.987**			
	Precision	Recall	F1	
Ears	1	1	1	
Beard	1	0.956	0.977	
Plate	1	1	1	
Roller	0.974	1	0.987	
Macro average	**0.993**	**0.989**	**0.991**	

4 Discussion

The result of using the ANN presented in Sect. 3 is the best of all analyzed. Its construction was preceded by several other experiments related to modeling using deep learning.

Firstly, we tried a model with a Recurrent Neural Network using LSTM architecture [5] whose structure is presented in Table 4. The embedding layer was used as the first layer with a 0.2 dropout rate and 5000 input dimensions. This layer turns indexes into dense vectors of fixed size. The LSTM layer had an input length of 100 and the dropout rate for gates set to 0.2 with hyperbolic tangent (tanh) activation function. The dense layer was used with a sigmoid activation function.

The best result for the LSTM model was achieved when 70% of the dataset was trained, 15% validated and 15% tested. The training average accuracy that has been achieved was 0.4736 and validation accuracy 0.5875 (Fig. 6) while average training and validation loss was close to 0 (Fig. 7).

In the second experiment, we built a model that was trained on 90%, validated and tested each on 5% of the data set. The following results were obtained: the average accuracy and validation were 1 (Fig. 8) and the average and validation loss were 0 (Fig. 9).

This type of behavior is characteristic of the overtrained ANN. Therefore, the training population was reduced to circa 70% that allowed to achieve more reliable results.

The problem of overtrained ANN occurred also during tests with an experimental model based on extended hidden layers consisting of 40 layers. By using the iterative method of selecting the appropriate number of layers, we decided to use 22 mixed dense, activation and dropout layers that allowed achieving reliable accuracy lower than 1 and loss higher than 0 (Table 5).

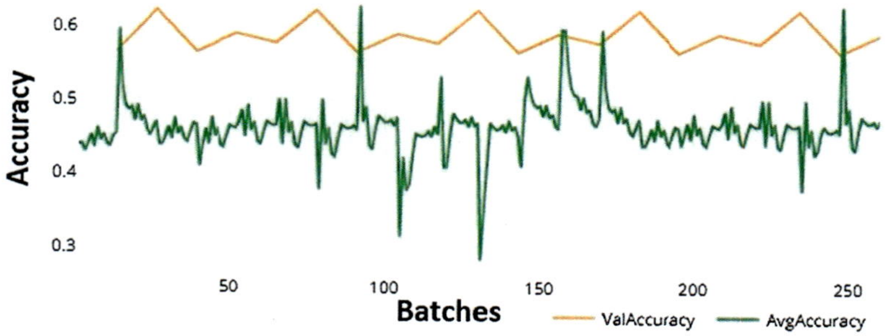

Fig. 6. The average accuracy (AvgAccuracy) and validation accuracy (ValAccuracy) for the LSTM model.

Table 4. Scheme of the ANN layers of the experimental model with LSTM layers.

No	Layer	Output dimensions
–	Input	700
1	Embedding	128
2	LSTM	128
3	Dense	4
–	Output	4

As a result, the average validation accuracy was 0.5104 (Fig. 10) and was not satisfactory. That model was improved by reducing it to 11 layers in total and adding two instances of the batch normalization layer (Table 2).

We considered the obtained model to be sufficient in terms of achieved accuracy and loss. Further modifications to the model would allow obtaining slightly better results than those already achieved, therefore we decided to use it as the final ANN model in our research, as presented in Sect. 2.3.

The chosen model was also tested in K-Fold Cross Validation, where K parameter was set to 10. Randomly selected 400 samples were used for training, 50 for validation and 200 for tests. Other parameters and hyperparameters remained unchanged. The average train accuracy that was achieved during these tests was 0.9696 with loss of 4.3602 and the validation average accuracy was 0.8406 with the loss of 0.3286. The obtained accuracies and losses are shown in Fig. 11.

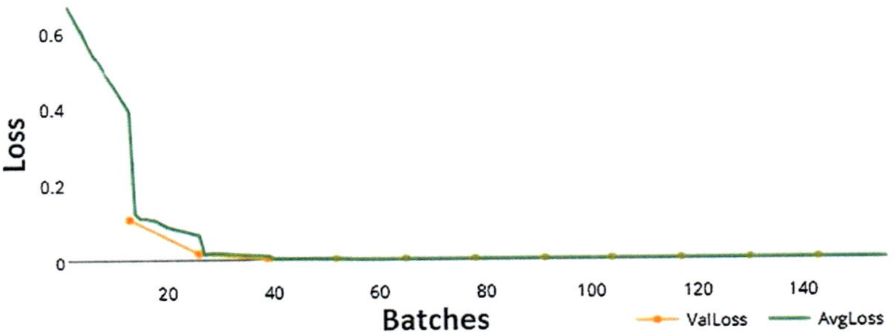

Fig. 7. The average loss (AvgLoss) and validation loss (ValLoss) for the LSTM model.

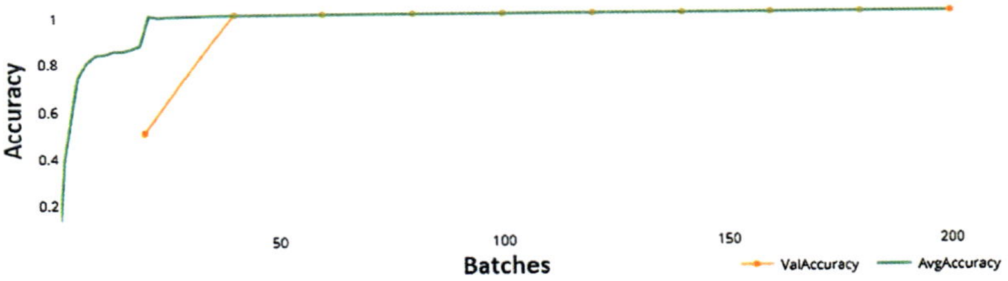

Fig. 8. The average accuracy (AvgAccuracy) and validation accuracy (ValAccuracy) for training with 90% of the data set.

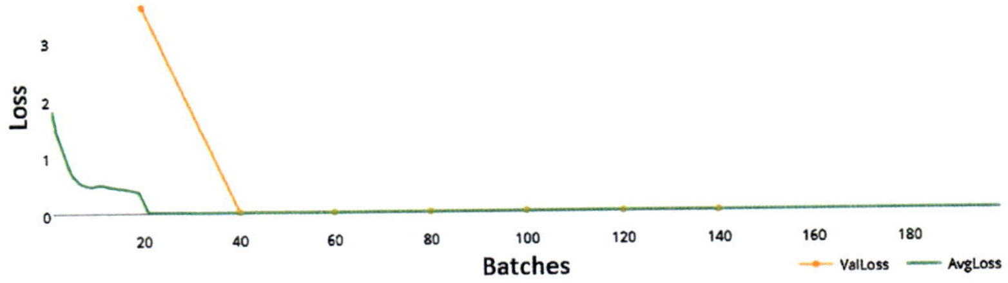

Fig. 9. The average loss (AvgLoss) and validation loss (ValLoss) for training with 90% of the data set.

Table 5. Scheme of the ANN layers of the experimental model with 22 hidden layers.

No	Layer	Output dimensions
–	Input	700
1	Dense	700
2	Activation	700
3	Dense	700
4	Activation	700
5	Dropout	700
6	Dense	700
7	Activation	700
8	Dense	700
9	Activation	700
10	Dropout	700
11	Dense	700
12	Activation	700
13	Dense	700
14	Activation	700
15	Dense	700
16	Activation	700
17	Dense	700
18	Activation	700
19	Dropout	700
20	Dense	700
21	Activation	700
22	Dense	4
–	Output	4

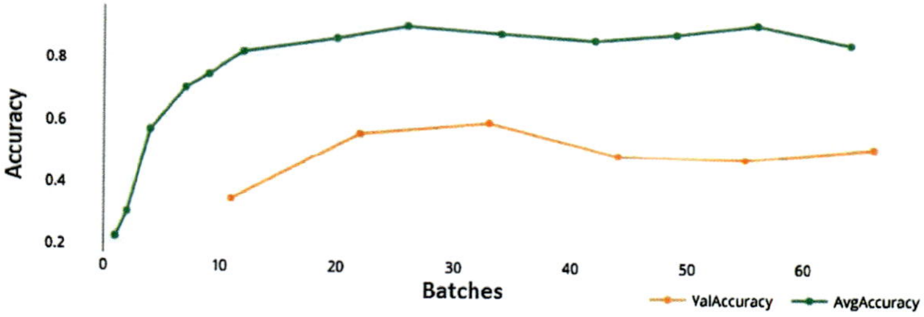

Fig. 10. The average accuracy (AvgAccuracy) and validation accuracy (ValAccuracy) for training with circa 70% of data set and 22 hidden layers.

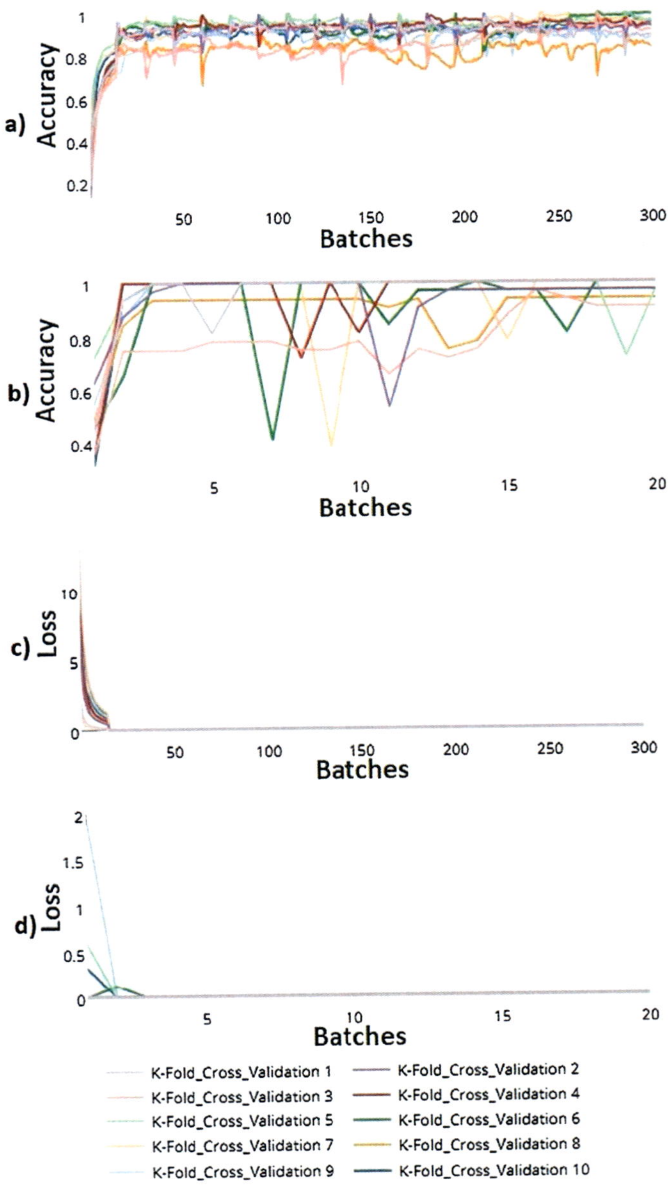

Fig. 11. The average: a) training accuracy, b) validation accuracy, c) training loss and d) validation loss for K-Fold Cross Validation.

5 Conclusions

The obtained results based on the experiments with various ANN models allowed us to choose the model that achieves high accuracy with low loss.

This research confirms that it is possible to recognize the flue pipe type basing on the spectrum. Therefore, it can also be assumed that an experienced listener, who knows well various organ pipe voices, can also correctly recognize the type of labium. This paper is also the incipience to further usage of deep learning methods in the area of a pipe organ.

Acknowledgements. Special thanks to the organbuilder Władysław Cepka for his invaluable help and providing the workshop and organ pipes for sound recording.

References

1. Angster, J., Rusz, P., Miklos, A.: Acoustics of organ pipes and future trends in the research. Acoust. Today **1**(13), 10–18 (2017)
2. Außerlechner, H., Trommer, T., Angster, J., Miklos, A.: Experimental jet velocity and edge tone investigations on a foot model of an organ pipe. J. Acoust. Soc. Am. **2**(126), 878–886 (2009). https://doi.org/10.1121/1.3158935
3. Deep Cognition Homepage. https://deepcognition.ai. Accessed 24 Apr 2020
4. Herremans, D., Chuan, C.: The emergence of deep learning: new opportunities for music and audio technologies. Neural Comput. Appl. **32**, 913–914 (2020). https://doi.org/10.1007/s00521-019-04166-0
5. Hochreiter, S., Schmidhuber, J.: Long short-term memory. Neural Comput. **9**(8), 1735–1780 (1997). https://doi.org/10.1162/neco.1997.9.8.1735
6. Hruška, V., Dlask, P.: Investigation of the sound source regions in open and closed organ pipes. Arch. Acoust. **3**(44), 467–474 (2019). https://doi.org/10.24425/aoa.2019.129262
7. Hruška, V., Dlask, P.: Connections between organ pipe noise and Shannon entropy of the airflow: preliminary results. Acta Acustica United Acustica **103**, 1100–1105 (2017)
8. Humphrey, E.J., Bello, J.P., LeCun, Y.: Feature learning and deep architectures: new directions for music informatics. J. Intell. Inf. Syst. **41**, 461–481 (2013). https://doi.org/10.1007/s10844-013-0248-5
9. Koutini, K., Chowdhury, S., Haunschmid, V., Eghbal-zadeh H., Widmer, G.: Emotion and theme recognition in music with frequency-aware RF-regularized CNNs. MediaEval 1919, 27–29 October 2019. ArXiv abs/1911.05833. Sophia Antipolis (2019)
10. Krizhevsky, A., Sutskever, I., Hinton, G.: ImageNet classification with deep convolutional neural networks. In: Advances in Neural Information Processing Systems, pp. 1097–1105 (2012). https://doi.org/10.1145/3065386
11. Lathi, B.P.: Linear systems and signals, 2nd edn. Oxford University Press, New York (2010)
12. Lehner, B., Widmer, G., Bock., S.: A low-latency, real-time-capable singing voice detection method with LSTM recurrent neural networks. In: 2015 23rd European Signal Processing Conference (EUSIPCO), Nice, pp. 21 – 25 (2015). https://doi.org/10.1109/EUSIPCO.2015.7362337

13. Rucz, P., Augusztinovicz, F., Angster, J., Preukschat, T., Miklos, A.: Acoustic behaviour of tuning slots of labial organ pipes.J. Acoust. Soc. Am. **5**(135), 3056–3065 (2014). https://doi.org/10.1121/1.4869679
14. Sakamoto, Y., Yoshikawa, S., Angster, J.: Acoustical investigations on the ears of flue or-GAN pipes. In: Forum Acusticum, pp. 647-651. EAA-Opakfi Hungary, Budapest (2005)
15. Srivastava, N., Hinton, G., Krizhevsky, A., Sutskever, I., Salakhutdinov, R.: Dropout: a simple way to prevent neural networks from overfitting. J. Mach. Learn. Res. **15**, 1929–1958 (2014). https://doi.org/10.5555/2627435.2670313
16. Vaik, I., Paal, G.: Flow simulations on an organ pipe foot model. J. Acoust. Soc. Am. **2**(133), 1102–1110 (2013). https://doi.org/10.1121/1.4773861
17. Verge, M., Fabre, B., Mahu, W., Hirschberg, A., et al.: Jet formation and jet velocity fluctuations in a flue organ pipe. J. Acoust. Soc. Am. **2**(95), 1119–1132 (1994). https://doi.org/10.1121/1.408460
18. Widmer, G.: On the potential of machine learning for music research. In: Miranda, E.R. (ed.) Readings in Music and Artificial Intelligence. Routledge, New York (2013)
19. Zeiler, M.: ADADELTA: an adaptive learning rate method. https://arxiv.org/abs/1212.5701. Accessed 24 Apr 2020

Industrial Applications

Adaptive Autonomous Machines - Modeling and Architecture

Lothar Hotz, Rainer Herzog, and Stephanie von Riegen(✉)

HITeC e.V., University of Hamburg, Hamburg, Germany
{hotz,herzog,svriegen}@informatik.uni-hamburg.de

Abstract. One of the challenges in mechanical and plant engineering is to adapt a plant to changing requirements or operating conditions at the plant operator's premises. Changes to the plants and their configuration require a well-coordinated cooperation with the machine manufacturer (or plant manufacturer in case of several machines) and, if necessary, with his suppliers, which requires a high effort due to the communication and delivery channels. An autonomous acting machine or component, which suggests and, if necessary, makes necessary changes by automatically triggered adjustments, would facilitate this process. In this paper, subtasks for the design of autonomous adaptive machines are identified and discussed. The underlying assumption is that changes of machines and components can be supported by configuration technologies, because these technologies handle variability and updates through automatic derivation methods, which calculate necessary changes of machines and components. A first architecture is presented, which takes into account the Asset Administration Shell (AAS) of the German Industry 4.0 initiative. Furthermore, three application scenarios are discussed.

Keywords: Knowledge representation · Configuration · Constraints · Ontology · Manufacturing systems

1 Introduction and Motivation

In recent years, the demand for the industrial production of small quantities has increased steadily. Whereas in the past, larger industrial plants were designed for the production of large quantities of exactly one product whose parameters did not change, today the possibility of fast, flexible adaptation to changes in product lines is becoming increasingly important, especially, if small lot sizes have to be processed. This development towards a more flexible and dynamic production is known as one of the key aspects of *Industry 4.0* (I4.0). While an adjustment of the machine settings is often sufficient for minor changes, larger adjustments require a modification of a machine by the machine manufacturer, or

This work has been developed in the project ADAM. ADAM (reference number: 01IS18077A) is partly funded by the German ministry of education and research (BMBF) within the research program ICT 2020.

M. Stettinger et al. (Eds.): ISMIS 2020, SCI 949, pp. 97–106, 2021.
https://doi.org/10.1007/978-3-030-67148-8_8

even changes of a complete production plant. For this purpose, the dependencies of individual plant components must be taken into account, e.g., the use of a stronger motor would possibly also require the use of a drive shaft that can withstand higher torques. If individual plant modules can be configured to give a higher or lower speed, instead of a higher throughput other modules could be enabled to achieve a higher accuracy.

The current adaptation process in plant engineering is depicted in Fig. 1. The plant operator recognizes from different triggers that the existing, running plant (1) must be changed. Trigger for this can be the availability of a higher performance component for the plant, changes in requirements for the machine, such as the desire to manufacture products from new materials, an increased throughput, or malfunctioning of a component (2). Updates in plants demand for multiple communications between the plant operator and the machine builder, who in turn may need to contact the machine-component manufacturer (3) to plan the adjustments. Once this process is complete, the adaptation of the plant is planned (4) and carried out (5). The adapted machine is then put back into operation (6).

If plants were to offer adaptation possibilities on their own initiative, the effort of this process could be significantly simplified for the plant operator and the parties involved in the adaptation process.

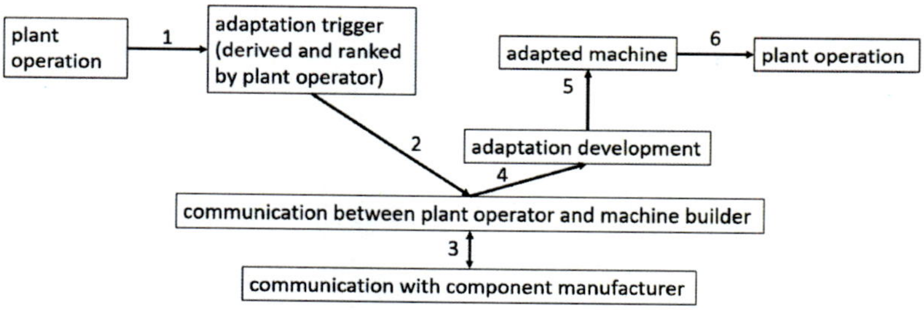

Fig. 1. Current adaptation process in plant engineering

In this paper, we present an innovative approach to adaptation planning for manufacturing plant processes. As our framework is intended to consider the RAMI 4.0 specifications [14] and taking profit from the standardisations covered by the AAS [13], these specifications will be briefly introduced in the next chapter, together with an overview of some related work. In Sect. 3, we present our concept of the autonomous agent. This concept is further addressed from the perspective of our architecture in Sect. 4. Section 5 presents three scenarios and shows different kinds of triggers to start the agent. In Sect. 6, we identify some technologies for realizing adapting machines and Sect. 7 summarizes the paper.

2 Related Work

One essential parameter towards a broad application of technologies is a clear and reliable standardisation of the relevant technologies, interfaces, and formats. Therefore, with the Reference Architecture Model Industry 4.0 (RAMI 4.0), a reference architecture model for a recurring situation was defined, which is intended to maintain a globally valid standard. It is designed as a cubic of layers, which facilitates a combined presentation of different aspects: It describes the architecture of technical objects (assets) and enables their description, the life cycle based on IEC 62890, and the assignment to technical or organisational hierarchies based on IECs 62264-1, 61512-1 [14].

In addition to the reference architecture model, all physical objects such as machine components, tools, factories but also products are combinedly represented with an Asset Administration Shell (AAS). This combination of each physical object with its AAS forms an Industry 4.0 Component. The AAS provides a minimal but sufficient description of an asset for exact identification and designation in its header part. The body part of the AAS consists of a number of independently maintained submodels. These represent different aspects of the relevant asset, i.e., properties and functions that can be used for different domains, such as a description regarding safety or efficiency, but could also outline various process capabilities. If the asset comprises I4.0 compliant communication infrastructure, it can be deployed directly to the asset, otherwise it is located in an affiliated IT system [13,19].

The development towards Industry 4.0 has been accompanied since decades by research [16], which has already developed partial solutions that address various aspects. Hoellthaler et al. designed a decision support system for factory operators and production planners. This system is intended to support them in responding appropriately on changing production requests by adding, changing or removing production resources. Based on optimization and material flow simulation, a result is computed that shows the best solution in terms of the highest number of parts produced and the lowest manufacturing costs per part, as well as alternative solutions [6]. Zhang et al. propose a five-dimensional model-driven reconfigurable Digital Twin (DT) to manage reconfiguration tasks and a virtual simulation to verify the applicability of system changes [18]. Contreras et al. demonstrate on the basis of a mixing station, which steps are necessary to design a RAMI 4.0 compatible manufacturing system [3]. Bougouffa et al. propose a concept that allows remote access using an Industry 4.0 interface on an open lab size automated production system [2]. Patzer et al. investigate the implementation of the AAS based on a specific use case with a clear focus on security analysis. Together with the description of their practical experiences, they provide recommendations for the implementation of the AAS on similar use cases [12]. In contrast to those approaches, we consider the use of knowledge-based configuration technologies [4], especially the configuration model describing the variants of a machine as well as the use of constraint programming for dealing with dependencies and relations, as a basic source for handling adaptations (see Sect. 3).

3 Concept for Autonomous Adapting Machines

We expect a machine to be accompanied by a complete description of the currently installed (probably parameterized) components, here called configuration. This can be a special submodel of the associated Asset Administration Shell. Each asset holds a number of constraints, which could be presented as part of a AAS submodel. An overview of the general concept for Autonomous Adapting Machines is given in [9]. The configuration is an instance of a configuration model, which is specified in machine-readable, semantically interpretable form [7]. It covers all the variants of system components. Due to the aspired standardisation process by the Industry 4.0 initiative, various machine component manufacturers and plant builders will be able to place their components in the configuration model. The standardisation and provision of components by several manufacturers in the configuration model alone can significantly reduce the effort of communication that would be necessary under traditional situations, if the configuration model is used as described below.

Furthermore, different types of constraints are stored in the configuration model as well as in the actual configuration itself (see Fig. 2). A component-related constraint could be, e.g., the maximum speed, torque, or the outer dimensions of a motor. Constraints can also describe the compatibility between several components. Other types of constraints are plant-related (e.g., maximum floor height), or production-related (minimum throughput). All these constraints will be taken into account by a constraint solver[1] whenever a reconfiguration process is executed.

A reconfiguration process can be activated by a trigger, which could be determined either continuously or event-driven. A trigger can be a sensor value (e.g., temperature, log entries), a new requirement of the plant operator to the asset (e.g., a customer-driven specification change) or an update of the configuration model. The update of the configuration model could be caused by a new component of the component manufacturer (e.g., provision of an optimized drive-system) or a new version of a firmware. If a trigger for adaptation occurs, the constraint solver determines whether the current configuration is sufficient to make the desired changes [8]. If this is not the case, the configuration model will be included in the process to reconfigure the asset. The result is the suggestion of one or more configurations. As the current configuration is also considered, possible solutions might be prioritized according to the fewest number of required changes. The plant operator can check and evaluate the results by applying them, if available, to the Digital Twin and might either immediately acknowledge the change (e.g., in case of the installation of a new firmware) or the proposed solution might require additional development activities to carry out the change. If no Digital Twin is accessible the evaluation has to be done by manually creating appropriate simulations. Due to its engineering knowledge model, the autonomous agent can participate in the verification done by the

[1] See, e.g., the Choco solver https://choco-solver.org/.

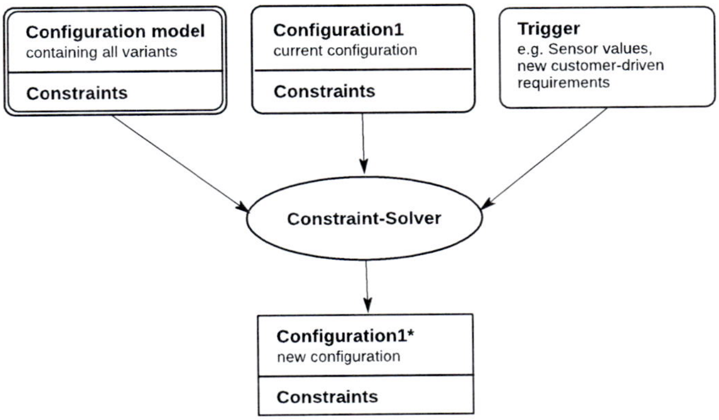

Fig. 2. Given the configuration model, a current configuration *Configuration 1*, and a trigger, a constraint solver computes an adapted configuration *Configuration 1** as input for the adaptation.

developer, for example, by checking the consistency of the solution, simulating processes, and evaluating predictions of the behavior after the changes.

By using simulation mechanisms provided by the DT, risks during implementation are reduced by monitoring the preparation steps and the consistency of parameters. In addition to simulation and verification, undesired emergence, which could arise from autonomous decisions, is recognized and ultimately prevented by monitoring. For this task, knowledge-based monitoring (based on [1]) monitors the activities of the autonomous agent. For this purpose, knowledge modeling about the possible adaptation activities of the agent as well as of the machine and its environment is used. This makes it possible to analyze and reflect on actions while the agent is performing them and, thus, to recognize unsafe actions and interactions. The simulation shows on the other hand the adapted behavior of the system using the simulation model, which processes the changes of the system behavior.

4 Architecture

Figure 3 shows the basic architecture of our Autonomous Adapting Machine (ADAM). The left side presents the structural model, which comprises the adaptation model as well as the machine description. From the view of the architecture, the adaptation model holds explicit knowledge of how to adapt the machine if some trigger is fulfilled. The current configuration of the asset is referenced here as "machine description". The structural model also contains the initial requirements for the asset. The "Trigger", depicted in Fig. 2, is fed into the system via the "process data". The process data contains, e.g., specific order data, which might pose requirements on the asset which were out of the scope

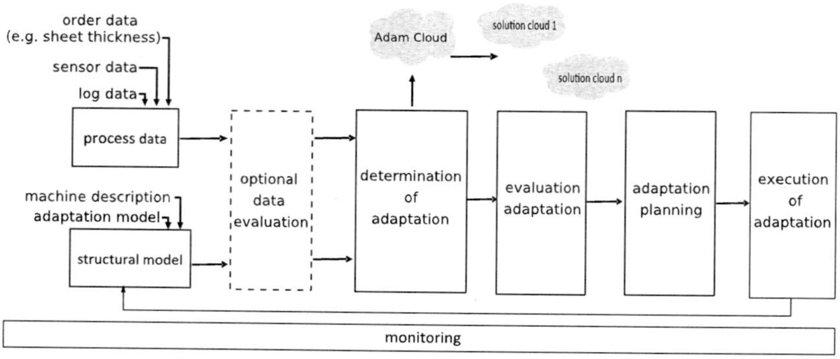

Fig. 3. Architecture of adaptive autonomous machines

of the initial requirements. Also sensor data and log data is part of the process data. Sensor data could include measured environmental conditions, but also disturbance variables such as an abnormally high power consumption of a motor identified by a threshold. The log information of the system will inform about the previous workload of the asset, but also, e.g., how often certain noncritical errors showed up within a specific time-span. These run-time parameters contained in the process data could provide a first indication that the current condition of the asset is inadequate, but it might also suggest that everything seems to be fine. The optional data evaluation then might filter out irrelevant or anonymize too sensitive data, which might contain intellectual property of the asset operator. This will partly be in the responsibility of the asset holder. The next part is named "determination of adaptation". Here, the above mentioned constraint solver will come into action. If no connection to cloud services is available, the solver will try to find a suitable solution based solely on the data of the adaptation model. However, the more advanced scenario is made possible by access to the "Adam Cloud". By transmitting the structural model and the treated process data to the Adam Cloud, which will have access to product services of asset or component manufacturers ("solution clouds"), a more advanced configuration model is build up, and optimized solution candidates will be returned. By the evaluation of process data of several assets, the Adam Cloud might also be able to provide some abnormality detection based on big data analysis, which could, e.g., improve the product life-cycle management of the asset. As described in the previous chapter, the returned solution candidates will be evaluated in a next step by simulation on a DT ("evaluation adaptation"). The suggestion of a possible adaptation, accompanied by a human expert, will then be put into the planning stage ("adaptation planning"), where additional outer conditions will be determined, which are necessary to execute the adaptation (e.g., the need or availability of human experts to change a component). After the adaptation is done, the new machine description and an updated adaptation model is written back to the structural model. The whole process is accompanied by the aforementioned "monitoring component".

5 Application Scenarios

We surveyed the challenges in current adaptation processes in collaboration with a component and plant manufacturer. The derived use cases can be mapped to one of the following representative scenarios.

In the first scenario, a new or optimized component, such as a more energy efficient drive, is offered by the manufacturer. This leads to a changed configuration model. Periodically the configuration model is queried. The constraint solver verifies whether the asset, described by the current configuration, is affected by the new component. The constraint solver then recommends a new configuration, if an improvement of the asset performance is achievable.

In the second scenario, a customer-driven change request must be considered, such as a new type of material of the desired product. The change request might lead to a new configuration determined by the constraint solver, but does not necessarily entail a change in the configuration model, namely in the case, where the configuration model already covers the new configuration.

The third scenario describes for example a sensor reporting faults in system operation, such as sheet metal plates that cannot be separated, which might be caused by higher humidity. These errors lead to log entries which are continuously evaluated by the agent and trigger the constraint solver. As far as the configuration model contains machine components that are able to prevail these faults, the constraint solver suggests a different configuration. Also in this third case, the configuration model is not changed.

6 Technologies for Creating Autonomous Adapting Machines

We identify following technologies for realizing autonomous adapting machines. Figure 4 depicts a summary of the proposed knowledge types. The configuration model of a machine represents all variants of the machine and its components [4]. The configuration model (depicted as *CM-C*) is distributed, i.e., the autonomous agent contains one part (CM_A-C) of the configuration model and the cloud of the component manufacturer another part (CM_C-C). CM_A-C contains the variants that existed during the time the machine was manufactured. It is updated if the machine is adapted. Considering that only some components supplied by a component manufacturer constitute a machine, CM_A-C has to be extracted from the configuration model that represents all components of a manufacturer. CM_C-C changes over time if the component manufacturer develops new components. Besides the configuration model the autonomous agent contains the actual configuration of the machine (current running hardware and software of the plant), i.e., an instance *CM-I* of the configuration model. Besides the configuration model *CM-C*, a requirement model *RM* will describe all possible requirements the components of *CM-C* shall supply [11]. Additionally to the requirements and the configuration model, we consider here a sensor model as a further artifact for structuring the knowledge of an autonomous agent. The sensor model *SM* represents all sensors that can acquire values about states in the environment [5, 10]. This model also entails knowledge about thresholds for deriving qualitative values about the world external to the machine. Those are mapped to the *RM* for deriving possible requirements *R* the machine has to fulfill [10, 11].

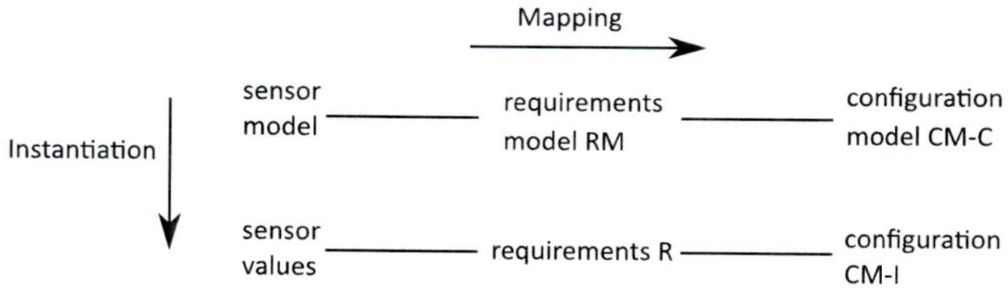

Fig. 4. Separation of models for sensor, requirements, and component knowledge in general (upper row) and for one machine (lower row)

By representing all those models and mappings as well as the identified new sensor values in a reasoning tool, a new configuration can be inferred with commonly known technologies [7]. Monitoring and verification of intended adaptations are further tasks which will apply simulation technologies and high-level monitoring of (here intended) activities [1]. The needed adaptations (e.g., component changes or updates) have to be identified, e.g., by comparing the original configuration and the adapted configuration. Furthermore, necessary planning actions have to be derived from a planning domain and finally executed [15]. All those technologies have to be combined in an architecture for autonomous adapting machines which includes decision about local and remote computations [15].

In a later step of our research, a further challenge will come into play, when interactions with other machines that become part of a collaborative system (as part of a changed manufacturing process) are considered. Although these cyber-physical systems are also considered under the term Industry 4.0, their focus is on the automatic setup of these systems for production. If adaptive systems are considered for cyber-physical systems, their adaptation must be considered as a further challenge. In the field of Internet of Things (IoT), similarly, the processing of sensor data is considered. Their combination with configuration tasks were also discussed by others, e.g., [5, 17].

The here used knowledge-based configuration or reconfiguration technologies are classic AI technologies. A constraint solver provides solutions based on a rule set, hence, it is possible to explain which constraints would be unsatisfactory in case of invalid solutions. This is a great advantage over the answers generated by systems based on, e.g., neural networks, which cannot be explained in this level of detail.

7 Summary

In this paper, we described the current situation in plant engineering. To support the needs reflected by Industry 4.0, we propose the use of configuration technologies not only in the beginning of a product life-cycle, but also during run-time of machines in production, also called reconfiguration. Knowledge about variants and dependencies, as well as reasoning methods known from the area of knowledge-based configuration can support the adaptation of machines. However, additional technologies, such as sensor evaluation, as well as adaptation planning, monitoring, and simulation on the basis of DT have to be considered. Knowledge-based configuration or reconfiguration can be classified as methods of classical AI-technologies, and a constraint solver provides solutions based on a simple or more advanced rule-set, still it is possible to show and therefore explain in detail, which constraints would be unsatisfied in case of invalid solutions. This results in an great advantage compared to the answers generated by neural network-based systems, which are not explainable in this level of detail. During our research, we identified concrete application scenarios for guiding the research in the direction of autonomous adaptive machines. As next steps, we consider the adaptation of the knowledge base to the specifications of AAS and RAMI 4.0 as well as the implementation of the architecture.

References

1. Bohlken, W., Koopmann, P., Hotz, L., Neumann, B.: Towards ontology-based real-time behaviour interpretation. In: Guesgen, H., Marsland, S. (eds.) Human Behavior Recognition Technologies: Intelligent Applications for Monitoring and Security, pp. 33–64. IGI Global (2013)
2. Bougouffa, S., Meßmer, K., Cha, S., Trunzer, E., Vogel-Heuser, B.: Industry 4.0 interface for dynamic reconfiguration of an open lab size automated production system to allow remote community experiments. In: IEEE International Conference on Industrial Engineering and Engineering Management, pp. 2058 – 2062 (2017). https://doi.org/10.1109/IEEM.2017.8290254
3. Contreras, J.D., Garcia, J.I., Pastrana, J.D.: Developing of industry 4.0 applications. International Journal of Online and Biomedical Engineering (iJOE) **13**(10), 30 – 47 (2017). 10.3991/ijoe.v13i10.733
4. Felfernig, A., Hotz, L., Bagley, C., Tiihonen, J.: Knowledge-Based Configuration: From Research to Business Cases. Morgan Kaufmann Publishers, Massachusetts (2014)
5. Felfernig, A., Falkner, A., Atas, M., Erdeniz, S.P., Uran, C., Azzoni, P.: ASP-based knowledge representations for IoT configuration scenarios. In: Proceedings of of the 19th Configuration Workshop, Paris, France, pp. 62 – 67, September 2017

6. Hoellthaler, G., et al.: Reconfiguration of production systems using optimization and material flow simulation. Procedia CIRP **81**, 133 – 138 (2019). https://doi.org/10.1016/j.procir.2019.03.024. 52nd CIRP Conference on Manufacturing Systems (CMS), Ljubljana, Slovenia, June 12-14, 2019
7. Hotz, L., Felfernig, A., Stumptner, M., Ryabokon, A., Bagley, C., Wolter, K.: Configuration knowledge representation & reasoning. In: Felfernig, A., Hotz, L., Bagley, C., Tiihonen, J. (eds.) Knowledge-Based Configuration – From Research to Business Cases, chap. 6, pp. 59–96. Morgan Kaufmann Publishers (2013)
8. Hotz, L., von Riegen, S., Herzog, R., Pein, R.: Towards a modular distributed configuration model for autonomous machines. In: Forza, C., Hvam, L., Felfernig, A. (eds.) Proceedings of the 22th Configuration Workshop, pp. 53–56. Università degli Studi di Padova, Italy, September 2020
9. Hotz, L., von Riegen, S., Herzog, R., Riebisch, M., Kiele-Dunsche, M.: Adaptive autonomous machines – requirements and challenges. In: Hotz, L., Krebs, T., Aldanondo, M. (eds.) Proceedings of of the 21th Configuration Workshop, pp. 61–64, September 2019
10. Hotz, L., Wolter, K.: Smarthome configuration model. In: Felfernig, A., Hotz, L., Bagley, C., Tiihonen, J. (eds.) Knowledge-Based Configuration – From Research to Business Cases, chap. 10, pp. 157–174. Morgan Kaufmann Publishers (2013)
11. Hotz, L., Wolter, K., Krebs, T., Deelstra, S., Sinnema, M., Nijhuis, J., MacGregor, J.: Configuration in Industrial Product Families - The ConIPF Methodology. IOS Press, Berlin (2006)
12. Patzer, F., Volz, F., Usländer, T., Blöcher, I., Beyerer, J.: The industrie 4.0 asset administration shell as information source for security analysis. In: IEEE International Conference on Emerging Technologies and Factory Automation, pp. 420 – 427 (2019). https://doi.org/10.1109/ETFA.2019.8869059
13. Plattform Industrie 4.0: Details of the Asset Administration Shell. https://www.plattform-i40.de/PI40/Redaktion/EN/Downloads/Publikation/Details-of-the-Asset-Administration-Shell-Part1.pdf
14. DIN SPEC 91345: Reference Architecture Model Industrie 4.0 (RAMI4.0). Beuth Verlag GmbH, Berlin, April 2016
15. Rockel, S., et al.: An ontology-based multi-level robot architecture for learning from experiences. In: Designing Intelligent Robots: Reintegrating AI II, AAAI Spring Symposium, Stanford, USA, pp. 52 – 57, March 2013
16. Scholz-Reiter, B., Freitag, M.: Autonomous processes in assembly systems. CIRP Ann. **56**(2), 712–729 (2007). https://doi.org/10.1016/j.cirp.2007.10.002
17. Schreiber, D., P.C., G., Lachmayer, R.: Modeling and configuration for Product-Service Systems: state of the art and future research. In: Proceedings of the 19th Configuration Workshop, Paris, France, pp. 72 – 79, September 2017
18. Zhang, C., Xu, W., Liu, J., Liu, Z., Zhou, Z., Pham, D.T.: A reconfigurable modeling approach for digital twin-based manufacturing system. Procedia CIRP **83**, 118–125 (2019). https://doi.org/10.1016/j.procir.2019.03.141. 11th CIRP Conference on Industrial Product-Service Systems
19. ZVEI e.V.: Struktur der Verwaltungsschale - Version 2, Fortentwicklung des Referenzmodells für die Industrie 4.0 - Komponente (2015). (in German)

Automated Completion of Partial Configurations as a Diagnosis Task Using FastDiag to Improve Performance

Cristian Vidal-Silva[1(✉)], José A. Galindo[2], Jesús Giráldez-Cru[3], and David Benavides[2]

[1] Departamento de Administración, Facultad de Economía y Administración, Universidad Católica del Norte, Antofagasta, Chile
cristian.vidal@ucn.cl

[2] Departamento de Lenguajes y Sistemas Informáticos, Universidad de Sevilla, Sevilla, Spain
{jagalindo,benavides}@us.es

[3] Andalusian Research Institute DaSCI "Data Science and Computational Intelligence", Universidad de Granada, Granada, Spain
jgiraldez@ugr.es

Abstract. The completion of partial configurations might represent an expensive computational task. Existing solutions, such as those which use modern constraint satisfaction solvers, perform a complete search, making them unsuitable on large-scale configurations. In this work, we propose an approach to define the completion of a partial configuration like a diagnosis task to solve it by applying the FastDiag algorithm, an efficient solution for preferred minimal diagnosis (updates) in the analyzed partial configuration. We evaluate our proposed method in the completion of partial configurations of random medium and large-size features models and the completion of partial configurations of a feature model of an adapted version of the Ubuntu Xenial OS. Our experimental analysis shows remarkable improvements in our solution regarding the use of classical CSP-based approaches for the same tasks.

Keywords: Partial configuration · Completion · FastDiag

1 Introduction

Configuration technology is a successful application of artificial intelligence (AI) [9]. Configuration technology permit notably reducing the development and maintenance costs of critical functionalities (features) that enables the mass customization realization [11]. Software product line (SPL) is an application domain for the mass-customization of software products [8].

Software product line engineering (SPLE) promotes the mass customization of software products (configurations) by identifying common and reusable features (e.g., functionalities) for the satisfaction of individual consumer requirements, and taking advantage of using a defined production framework [1,6].

M. Stettinger et al. (Eds.): ISMIS 2020, SCI 949, pp. 107–117, 2021.
https://doi.org/10.1007/978-3-030-67148-8_9

SPLE relies on efficient mechanisms to detect and diagnose anomalies in configurations, i.e., finding configurations that violate some constraints of the SPL and explaining the reasons for such inconsistency. To this purpose, feature models (FMs) have been proposed as a compact abstraction of families of products. FMs allow to represent all the existing features in the family and constraints among them [17], and they represent the valid product configurations of a software product family [6]. Although in this work we consider basic FMs [4], our proposed solution can be directly applied to more complex FMs, such as cardinality-based FMs [14] and attributed FMs [18], and on other configurations approaches supported by some reasoning technology. We use FMs as an example to illustrate our approach.

The completion of partial configurations consist of finding the set of non-selected components necessary for getting a complete configuration. In FM configurations, each feature is decided to be either present or absent in the resulting products, whereas in partial configurations, some features are undecided. The completion of partial configuration is a non-trivial and computationally expensive task due to the existence of constraints among features of FMs [15], and more expensive in large-scale FMs. Configurations can result in misconfigurations (i.e., non-valid configurations) which can impact on the system availability [25]. Known misconfiguration examples are the unavailability of Facebook platform [7], service-level problems of Google [3], and invalid operation of Hadoop clusters [19].

In the literature there exist efficient algorithms for the automated analysis and diagnosis of FMs, such as FastDiag and FlexDiag [8,11,12]. FastDiag and FlexDiag rely on encoding the FM constraints into the formal representation of reasoning technology for diagnosis in those configurations using off-the-shelf solvers (e.g. Constraint Satisfaction Problem and SATisfiability Problems, that is, CSP and SAT, respectively). Existing computer-assisted methods for the completion of partial configurations, such as modern CSP solvers, often apply computationally expensive complete search functions [24]. Hence, to find a consistent FM configuration of FMs with n features requires exploring 2^n possible configurations in the worst case. Moreover, CSP and SAT solver solutions can be minimal, but they not always represent the preferred configuration [22].

In this work, we define the completion of partial configuration as diagnosis task to solve it by using an efficient diagnosis algorithm. Specifically, we use the diagnosis algorithm FastDiag to evaluate its performance regarding applying a traditional CSP-based approach. Our experiment consists of random products of a set of FMs, both randomly generated by the Betty toolkit, and partial configurations of the FM of a Ubuntu Xenial version. The obtained results show that our proposal is several orders of magnitude faster than the applied traditional CSP-based approach. Thus, Our contributions are the following:

- We define the completion of partial configurations as a diagnosis task. This approach allows us to directly apply FastDiag to get a consistent configuration from a predefined set of features (i.e., from a partial configuration).

- We provide a public available implementation of our solution in the FaMa platform [5], as well as a set of available models and configurations.

The rest of the article is organized as follows. Section 2 describes preliminary background on FM and existing diagnosis solutions. Section 3 establishes the background to define the completion of partial configurations as diagnosis problems. Section 4 defines case studies and presents the application results of our solution. Some related works are described in Sect. 5. Finally, Sect. 6 concludes and proposes future work.

2 Preliminaries

2.1 Feature Models and Completion of Partial Configurations

An FM is a tree-like hierarchical representation of features and their constraints for a family of products. The following types of constraints exist in basic FMs: (a) parent-children or inclusion relationships, and (b) cross-tree constraints (CTC). Four kinds of inclusion relationships exist: (i) *mandatory* (the parent requires its child, and *vice versa*), (ii) *optional* (the parent does not require its child), (iii) *inclusive*-OR (the parent requires at least one of the set of children), and (iv) *alternative*-XOR (the parent requires exactly one of the set of children). CTC of traditional FMs are (v) *requires* (a feature requires another), and (vi) *excludes* (two features cannot be in the same configuration). Figure 1 illustrates a FM with a root feature *Debian* that has the mandatory children *texteditor*, *bash* and *gui*. Feature *texteditor* has an inclusive set of children features *vi*, *gedit* and *openoffice.org-1*, and *openoffice.org-1* also has two optional children features *openoffice.org-1.1* and *openoffice.org-1.2*. Feature *gui* has an inclusive set of children features *gnome* and *kde*. Feature *gnome* is required by feature *openoffice.org-1*.

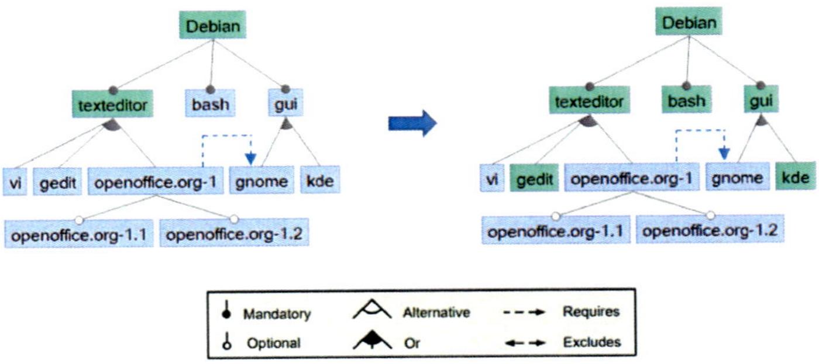

Fig. 1. An example of a partial configuration completion in the Debian FM.

Figure 1 illustrates this problem (the features in green represent selected features). The partial configuration in the left $\{Debian, texteditor\}$ is extended to the complete configuration $\{Debian, texteditor, gedit, bash, gui, kde\}$ in the right. We construct these models using FeatureIDE [20].

2.2 FM Configuration and Diagnosis Tasks

An FM configuration task (F, D, C) consists of setting the values of a set of features $F = \{f_1, \ldots, f_n\}$ in a common domain D to satisfy a set of configuration constraints $C = CF \cup CR$. CF represents the FM base knowledge (i.e., constraints among the features) and CR the user preferences (i.e., desired features in the product) [8], and D is {`true`, `false`} usually. Hence, a complete configuration represents a setting of each feature f_i in F respecting the configuration constraints of C.

We require diagnosis operations for identifying solutions for configurations that violate the FM constraints. For a consistent knowledge base AC, and a non-consistent configuration S, a diagnosis task (S, AC) gives a set of constraints or diagnosis $\Delta \subseteq S$ such that $(AC - \Delta)$ is consistent. Δ is minimal if $\neg\exists\Delta' \subset \Delta$ satisfying the diagnosis property in AC. S is the set of selected features in the configuration. The presence or absence of a features f_i can be expressed as $f_i =$ `true` or $f_i =$ `false`, respectively.

3 Minimal Completion of Configurations by Diagnosis

As was mentioned in the previous section, to proceed with an FM diagnosis, FastDiag receives the parameters S and AC, that is, the user preferences and the FM knowledge base that contains S. The completion of a partial configuration is a diagnosis task to find the preferred minimal set of features to select for getting a full configuration. Hence, the main task to apply FastDiag for diagnosis a preferred minimal completion is to define the knowledge base and the suspicious set of constraints in conflict.

An FM formally represents a set of features F and a set of constraints C. A partial configuration can be seen as a set of assigned features S, i.e., $S \subset F$. Likewise, we define the set of unassigned features nS as $nS = (F - S) \neq \emptyset$. The partial configuration S is valid if $C \cup S$ is consistent, which, for the sake of clarity, we always assume to hold. To find the remaining features for a complete configuration, we run FastDiag with $S = nS$, and a knowledge base $C \cup S$. We assign Boolean values to the components of S and nS for consistency checks in the FM. FastDiag returns a preferred minimal set $\Delta \subseteq nS$ of features necessary for the completion of the partial configuration S. In summary, we define a FastDiag application for diagnosis features for the completion of partial products. Table 1 gives our definition for that task. The sets S and nS represent the selected and non-selected features in the partial configuration p respectively, and C is the set of base constraints in the FM. Because FastDiag works on constraints in a reasoning solver tool such as CSP and SAT, our solution is not restricted to work only for the FMs completion. We suggest to read [10,13] for more details of FastDiag.

Table 1. Diagnosis-based solution for the completion of a partial configuration using FastDiag.

Analysis operation	Property check	Explanation (Diagnosis)
Completion of Partial Configurations	Diagnosis in nS (set of non-selected features)	$FastDiag(nS, C \cup S \cup nS)$

FastDiag gives an ordered by preference set of features to update using a lexicographical order by default. Our solution can use personalized options for selecting features such as randomly, the nearest feature to some already chosen feature, or based on a priority ranking regarding previous configurations.

4 Empirical Evaluation

To evaluate the performance of our solution, first, we generate a set of random FMs using the Betty tool-suite [23] to define the number of features, structure and the number of cross-tree constrain of randomly generated FMs. We generate models with the following number of features $|F| = \{50, 100, 500, 1000, 2000, 5000\}$ and with next percentages of CTC $c = \{5, 10, 30, 50, 100\}$. For each model, we generate partial configurations with the following percentage of assigned features $a = \{10, 30, 50, 100\}$. We generate 10 random instances for each model and partial configuration.

Table 2. Avg. time (in milliseconds) on the completion of partial configuration of randomly generated Betty FMs by the number of features n.

n	CSP-based app	FastDiag app	% speed-up
50	98.00	**97.90**	0.10
100	109.06	**04.63**	4.06
500	200.09	**71.60**	14.24
1,000	405.89	**258.26**	36.37
2,000	1,392.27	**411.01**	70.48
5,000	15,677.93	**808.61**	94.84
All	2,980.54	**308.67**	36,58

Our proposal is evaluated using FastDiag in the FaMa tool suite with the Choco CSP solver [5] for consistency checks. The CSP-based approach uses the same solver. In what follows, we report the results comparison of both approaches. In both results, best values are marked in bold. In Table 2 we report the average solving time of the CSP-based and the FastDiag approach, aggregating the results by the number of features in the random models. In Tables 3 and 4, we report the same results aggregated by the percentage of CTC and the percentage of feature in the partial configurations, respectively. The last column ($\%speed-up$) of each table shows the percentage of improvement of our solution regarding the CSP-based approach. The last row presents the average results.

In each comparison of result, the FastDiag diagnosis solution is faster than the CSP-based approach. In general, there are noticeable differences, in some cases with a speed-up greater than 19x (see $n = 5000$ in Table 2). Hence, the performance improvements are bigger as the number of features in the FM increase. This is possibly due to the complete search of the CSP solver that scales exponentially. In contrast, our solution seems to scale much better. Notice that the speed-ups in Table 2 increases for greater values of n. On the contrary, as Tables 3 and 4 show, the number of CTC and size of the partial configuration seem to have a low effect on the performance of both solutions. However, there are two remarkable effects. First, the speed-up of FastDiag slightly decreases as the number of CTC increases. This suggests that if the number of constraints is exponentially bigger, there may be cases in which both approaches perform similarly. Second, the speed-up of FastDiag slightly decrease as the partial configuration becomes smaller. This suggests that the larger the size of such a partial configuration, the bigger the differences between FastDiag and the CSP-based approach.

Table 3. Avg. time (in milliseconds) on the completion of partial configuration of randomly generated Betty FMs by the % of CTC cc.

c	CSP-based app	FastDiag app	% speed-up
5	2,953.98	**275.80**	90,67
10	2,993.21	**282.57**	90,56
30	2,971.61	**303.87**	89,77
50	2,937.65	**308.72**	89,49
100	3,046.24	**372.38**	87,78
All	2,980.54	**308.67**	89,64

Table 4. Avg, time (in milliseconds) on the completion of partial configuration of randomly generated Betty FMs by the % of features in the partial configuration a.

a	CSP-based app	FastDiag app	% speed-up
10	2,954.96	**318.23**	89.23
30	2,973.43	**315.03**	89,41
50	2,966.08	**306.58**	89.66
100	3,027.68	**294.82**	90,26
All	2,980.54	**308.67**	90,26

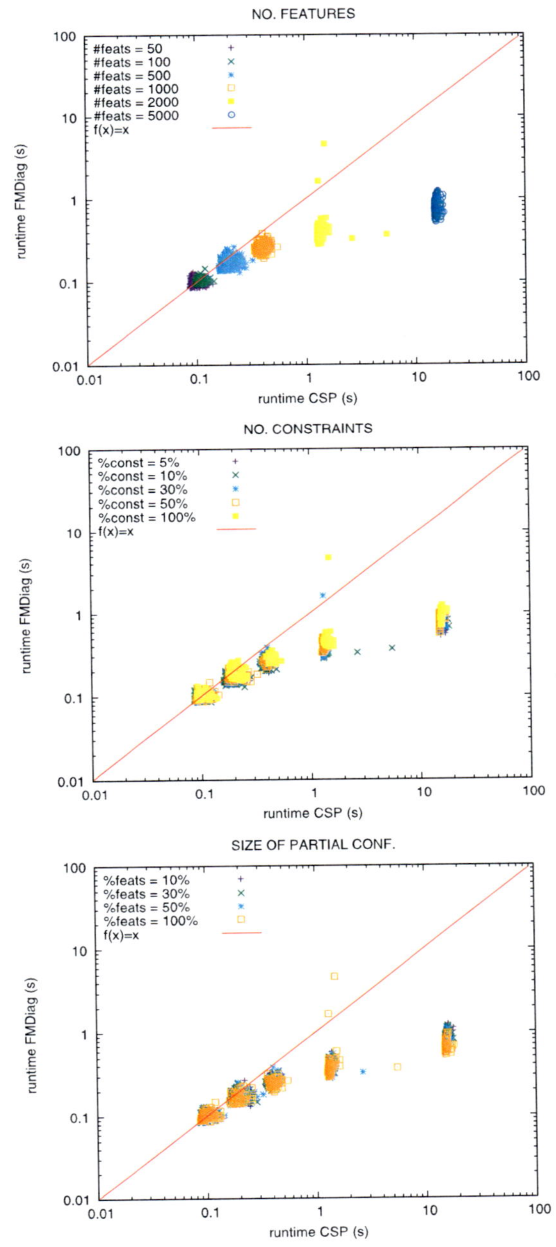

Fig. 2. Scatter plot of CSP-based versus FastDiag approach on the completion of partial configuration of randomly generated Betty FMs (in seconds), aggregated by the number of features (top left), by the percentage of CTC (top right), and by the percentage of features in the partial configuration (bottom).

Figure 2 shows the scatter plot of both approaches, i.e., the solving time of the CSP-based approach in the X axis versus the FastDiag approach in the Y axis. This plot confirms the expected performance from the aggregated results in previous tables. In particular, we can observe that the solution based on diagnosis algorithm scales quite well with the number of features of the generated model, whereas there are only small differences when this solution is compared with respect to other parameters, such as the number of CTC or the size of the partial configuration.

As a second performance evaluation, we generated an FM and partial products for the Ubuntu Xenial OS. We generate five valid partial products of 5%, 10%, and 15% of a complete and valid configuration for that FM, respectively. Then, we apply the CSP-based approach and FastDiag for the completion of those products. Table 5 and Fig. 3 show the computation results. Like in the previous report, speed-up percentages exist in these tests that confirm the efficiency of our solution.

All experiments were executed in an Intel(R) Core (TM) i7-3537U CPU @ 2.00 GHz with 4 GB RAM using a Windows 10 64 bits operating system.

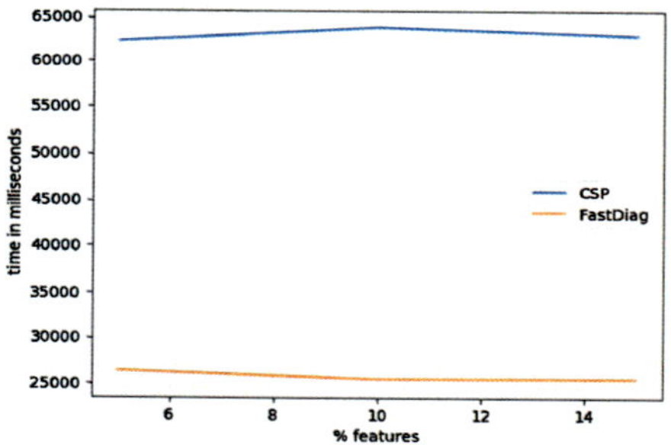

Fig. 3. Runtime execution for the completion of partial products for a FM of Ubuntu Xenial.

Table 5. Avg. solving time (in milliseconds) on the completion of partial configuration of randomly generated partial configuration of an FM for a reduced version of the Ubuntu Xenial OS by the percentage of features in the partial configuration.

$\%Features$	CSP-based app	FastDiag app	% speed-up
5	62,297.05	**26,318.14**	57,75
10	63,885.43	**25,467.03**	60,14
15	62,973.01	**25,508.02**	59,49
All	63,051.83	**25,764.40**	59,13

5 Related Work

Reiter [21] introduces the Hitting Set Directed Acyclic Graph for diagnosis using a breadth-first search on conflict sets. Bakker et al. [2] apply a model-based diagnosis to identify the set of relaxable constraints on conflict sets in a CSP context. Junker [16] proposes QuickXplain, a divide-and-conquer approach to significantly accelerate the conflict detection on over-constrained problems. Following the QuickXplain strategy, Felfernig et al. [10] present FastDiag for efficient diagnosis solutions on customer requirements in configuration knowledge bases. The work of [8] review and apply FastDiag on the FM diagnosis, and [11,12] describe FlexDiag, a FastDiag extension, for anytime diagnosis scenarios and apply both algorithms for the FMs diagnosis. Felfernig et al. [8] highlight an advantageous property of FastDiag regarding existing diagnosis approaches: FastDiag is a direct-diagnosis solution without a preceding conflict detection, that is, FastDiag uses a conflict-independent search strategy [9].

6 Conclusions

In this work, we defined a solution for the completion of partial configurations as a diagnosis problem that allows us to apply FastDiag to this problem. Experimental results with FMs show that this approach improve a traditional solution approach in several orders of magnitude, achieving speed-ups of more than 19x in some cases. Therefore, FastDiag also represents an efficient solution for the completion of configurations in software product lines.

We plan to adapt these ideas to FlexDiag in real-time scenarios with predefined time limits and acceptable trade-offs between diagnosis quality and efficiency of the diagnostic reasoning.

Acknowledgements. This work has een partially funded by the EU FEDER program, the MINECO project OPHELIA (RTI2018-101204-B-C22); the TASOVA network (MCIU-AEI TIN2017-90644-REDT); and the Junta de Andalucia METAMORFOSIS project.

References

1. Apel, S., Batory, D., Kstner, C., Saake, G.: Feature-Oriented Software Product Lines: Concepts and Implementation. Springer, Heidelberg (2013)
2. Bakker, R.R., Dikker, F., Tempelman, F., Wognum, P.M.: Diagnosing and solving over-determined constraint satisfaction problems, pp. 276–281 (1993)
3. Barroso, L.A., Hoelzle, U.: The Datacenter As a Computer: An Introduction to the Design of Warehouse-Scale Machines. Morgan and Claypool Publishers, 1st edn. (2009)
4. Batory, D.: Feature models, grammars, and propositional formulas. In: Proceedings of the 9th International Conference on Software Product Lines, pp. 7–20 (2005). https://doi.org/10.1007/11554844_3

5. Benavides, D., Segura, S., Trinidad, P., Ruiz–Cortés, A.: FAMA: tooling a framework for the automated analysis of feature models. In: Proceeding of the 1st International Workshop on Variability Modelling of Software-Intensive Systems (VAMOS), pp. 129–134 (2007)
6. Benavides, D., Segura, S., Ruiz-Cortés, A.: Automated analysis of feature models 20 years later: a literature review. J. Inf. Syst. **35**(6), 615–636 (2010)
7. Facebook: More details on today's outage. https://m.facebook.com/nt/screen/?params=%7B%22note_id%22%3A10158791436142200%7D&path=%2Fnotes%2F%7Bnote_id%7D&_rdr. Accessed 13 May 2018
8. Felfernig, A., Benavides, D., Galindo, J., Reinfrank, F.: Towards anomaly explanation in feature models. In: Proceedings of the 15th International Configuration Workshop (2013)
9. Felfernig, A., Hotz, L., Bagley, C., Tiihonen, J.: Knowledge-Based Configuration: From Research to Business Cases, 1st edn. Morgan Kaufmann Publishers Inc., San Francisco (2014)
10. Felfernig, A., Schubert, M., Zehentner, C.: An efficient diagnosis algorithm for inconsistent constraint sets. Artif. Intell. Eng. Design Anal. Manuf. **26**(1), 53–62 (2012)
11. Felfernig, A., Walter, R., Galindo, J.A., Benavides, D., Polat Erdeniz, S., Atas, M., Reiterer, S.: Anytime diagnosis for reconfiguration. J. Intell. Inf. Syst. **51**, 161–182 (2018)
12. Felfernig, A., Walter, R., Reiterer, S.: FlexDiag: anytime diagnosis for reconfiguration. In: Proceedings of the 17th International Configuration Workshop (2015)
13. Fernández-Amorós, D., Heradio, R., Cerrada, J.A., Cerrada, C.: Ascalable approach to exact model and commonality counting for extended feature models. IEEE Trans. Software Eng. **40**(9), 895–910 (2014). https://doi.org/10.1109/TSE.2014.2331073
14. Gómez, A., Ramos, I.: Automatic tool support for cardinality-based feature modeling with model constraints for information systems development. In: Information Systems Development, Business Systems and Services: Modeling and Development [Proceedings of ISD 2010, Charles University in Prague, Czech Republic, August 25-27, 2010], pp. 271–284 (2010). https://doi.org/10.1007/978-1-4419-9790-6_22
15. Ibraheem, S., Ghoul, S.: Software evolution: a features variability modeling approach. J. Softw. Eng. **11**, 12–21 (2017)
16. Junker, U.: QUICKXPLAIN: preferred explanations and relaxations for over-constrained problems. In: Proceedings of the 19th National Conference on Artificial Intelligence (AAAI), pp. 167–172 (2004)
17. Kang, K., Cohen, S., Hess, J., Novak, W., Peterson, A.: Feature-oriented domain analysis (FODA) feasibility study. Technical Report CMU/SEI-90-TR-021, Software Engineering Institute, Carnegie Mellon University (1990)
18. Karataş, A.S., Oğuztüzün, H., Doğru, A.: From extended feature models to constraint logic programming. Sci. Comput. Program. **78**(12), 2295 – 2312 (2013). https://doi.org/10.1016/j.scico.2012.06.004. http://www.sciencedirect.com/science/article/pii/S0167642312001153. Special Section on International Software Product Line Conference 2010 and Fundamentals of Software Engineering (selected papers of FSEN 2011)
19. Li, J.Z., et al.: Challenges to error diagnosis in Hadoop ecosystems. In: Proceedings of the 27th Large Installation System Administration Conference (LISA), pp. 145–154 (2013)
20. Meinicke, J., Thüm, T., Schröter, R., Benduhn, F., Leich, T., Saake, G.: Mastering Software Variability with FeatureIDE. Springer, Cham (2017)

21. Reiter, R.: A theory of diagnosis from first principles. AI J. **23**(1), 57–95 (1987)
22. Riener, H., Fey, G.: Exact diagnosis using Boolean satisfiability. In: Proceedings of the 35th International Conference on Computer-Aided Design. ICCAD 2016. Association for Computing Machinery, New York (2016). https://doi.org/10.1145/2966986.2967036
23. Segura, S., Galindo, J.A., Benavides, D., Parejo, J.A., Ruiz-Cortés, A.: BeTTy: benchmarking and testing on the automated analysis of feature models. In: Proceedings of the Sixth International Workshop on Variability Modeling of Software-Intensive Systems, pp. 63–71 (2012)
24. White, J., Benavides, D., Schmidt, D.C., Trinidad, P., Dougherty, B., Cortés, A.R.: Automated diagnosis of feature model configurations. J. Syst. Softw. **83**(7), 1094–1107 (2010)
25. Yin, Z., Ma, X., Zheng, J., Zhou, Y., Bairavasundaram, L.N., Pasupathy, S.: An empirical study on configuration errors in commercial and open source systems. In: Proceedings of the Twenty-Third ACM Symposium on Operating Systems Principles, pp. 159–172 (2011)

Exploring Configurator Users' Motivational Drivers for Digital Social Interaction

Chiara Grosso[1(✉)] and Cipriano Forza[2]

[1] Department of Management, University Cà Foscari Venice, Venice, Italy
chiara.grosso@unive.it
[2] Department of Management and Engineering, University of Padua, Vicenza, Italy

Abstract. At a global level, the demand for online transactions is increasing. This is propelled by both the digital transformation paradigm and the COVID 19 pandemic. The research on Web infrastructure design recognizes the impact that social, behavioral, and human aspects have on online transactions in e-commerce, e-health, e-education, and e-work. As a result, social computing features are leading the Web with information and communication technologies that facilitate interactions among web users through socially enhanced online environments. It is crucial to research the social, behavioral, and human dimensions of web-mediated activities, especially when social activities are restricted only to an online environment. The present study focuses on the social dimension of the e-commerce of customizable products. This domain was selected because of the specificity of its product self-design process in terms of customers' decision-making and their involvement in product value creation. This study aims to seek the extent that a set of customers' motivational drivers rely on their need to interact with real persons during the technology-assisted process of products' self-design. By adopting a user-centered perspective, the study considers 937 self-design experiences by 187 young adult users on a sample of 378 business-to-customers product configurators. The results should provide companies and software designers with insights about customers' need for social presence during their product self-design experience so that they can fulfill this need by using social technology that provides positive experiences.

Keywords: Online sales configurator · Social software · Social product configuration systems · User experience (UX)

1 Introduction

The digital transformation paradigm [1] and the current global health emergency require the business ecosystem to rapidly adjust its strategy to the evolution of web technology and infrastructures. This adjustment needs to be rapid for at least two reasons: (i) the worldwide demand for online transactions is increasing and (ii) web social technologies are facilitating and supporting interactions between web users with socially enhanced online environments. As a result, web social technologies that connect customers worldwide are changing the expectation that consumers have with online transactions in terms

M. Stettinger et al. (Eds.): ISMIS 2020, SCI 949, pp. 118–138, 2021.
https://doi.org/10.1007/978-3-030-67148-8_10

of social presence and social interactions. Social presence is defined in literature on computer-mediated communication as the capacity of a medium to provide its users the "feeling of being there with a 'real' person" ([2], p.1) to convey human contact and sociability.

As stated in previous research on the digital business ecosystem, companies that effectively manage digital technologies gain better customer experience, streamlined operations, and new business models [3].

Despite the recognition of the urgency for digital transformation strategies to respond to customers' new expectations, most companies lack the knowledge to drive transformation through web social technologies [3]. To reduce this gap, research is needed to investigate customers' new behaviors and their need for social interaction during their online transactions. This research should help companies design technology-assisted experiences that properly respond to customer expectations.

The present study moves a step forward in this direction by investigating customers' expectations in terms of digital social interaction in the specific domain of the e-commerce of customizable products. This domain was selected because its specific process self-design product involves customers in the decision-making and a different number of choice tasks is required before an optimal solution is produced. Thus, customers may need support for their decision-making process through contact with real persons in addition to the support provided by product configurators [4] and/or recommender systems [5, 6] and enabled by social technology features. The self-design of products provides customers with several benefits both in terms of experience [7] and possession of a customized product [8]. Thus, involving customers in product value creation can be a strategy to engage customers and to differentiate companies in online markets.

As stated in previous research [9], designing gratifying product customization experiences triggers positive responses among potential customers, which are carried over to the assessment of product value ([10], p. 1029). Rewarding the mass-customization experience is, therefore, one way to increase customers' willingness to pay for the self-designed product [10, 11]. As a result, mass customizers may increase their sales volumes as rewarding shopping experiences lead to higher repurchase intentions [12, 13].

The main question that the present study aims to answer is how to integrate social technology into self-design environments to make positive experiences (almost) certain for its users. To answer this more generic question, the determinants that trigger users' need for digital social interaction during their decision-making processes must be investigated. To this end, we explore a set of consumers' motivational drivers to seek to what extent they underlie users' need to digitally interact with real persons during their product self-design experience.

To perform the empirical exploration, we use data collected from a sample of 187 young adults who carried out 937 product self-design experiences (also referred to in this study as configuration experiences, product configuration, or configuration) on 378 online active business-to-customer (B2C) online sales configurators (OSCs) of different goods. The analysis considers each step of the users' product self-design process via online sales configurators. Results from the present study provide B2C companies that sell customizable products with insights on users/customers' needs for digital social interaction.

These insights can help companies understand how to manage social technology to fulfill their customers' expectations. They also help software designers understand how to reduce the possible mismatch between companies' e-commerce strategies and users' actual experiences, thus designing (almost) certainly positive experiences for their users/customers.

2 Related Works

The following sections provide a review of related works. They situate the contributions that the present study aims to provide in the domains of information systems, computer-mediated communication, product self-design process, and customers' behaviors.

2.1 Social Presence

Web social technologies are leading the online world by facilitating and supporting user interactions with web-based features of digital social interactions, such as creating, evaluating, and exchanging user-generated content [14]. The range of social technology-mediated interactions available for web users who shop online (now called digital customers) is now quite diverse. Examples of these are reviewing and rating products and collaborative shopping experiences that allow consumers to maintain high levels of control over their online transactions Huang 2015 [15]. Online environments, including e-shops, are increasingly enhancing their capacity to provide users with the "feeling of being there with a 'real' person" ([2], p. 1). The capacity of a medium to instill this feeling is defined in the literature on computer-mediated communication as social presence. Social presence is recognized as a crucial component of interactions that take place in virtual environments wherein individuals could coexist and interact with each other [16]. A medium can enable this feeling of "warmth" by incorporating one or more web-based features that allow users to interact with other humans such after-sales e-mail support [17], virtual communities, chat [18], message boards and human web assistants [19].

In online shopping, social presence is associated with a variety of positive communication outcomes, which lead to greater purchase intentions, such as trust, enjoyment, and perceived usefulness of an online shopping website [20]. Despite existing research on the B2C product customization process that has recognized the importance of social feedback and social interactivity between configurator users [21–24], the research on users' need to digitally interact with real persons is surprisingly still in its infancy. As a result, a growing number of product configurators have started to connect to social software that enables social interactive features. However, up to the date of the present study, none of the features integrated into configuration systems support users in selecting one or more communication partners on-demand whenever they look for proactive support at different steps of their decision-making process [25]. Moreover, results from previous studies on social product configuration systems are contradictory [23]. As an example, Franke et al. [26] found that integrating user communities into self-design processes increased user satisfaction, purchase intention, and willingness to pay. However,

Moreau and Herd [22] showed that social comparisons between configurator users can lower consumers' evaluations of their self-designed products. The state of the art in this area calls for more investigation on users' need for digital social interactions and their specificities.

To this end, the present study explores to what extent a set of motivational drivers underlie users' intention to interact with one or more communication partners (such as personal contacts, experts from the company and other configurator users) to be supported at each step of their configuration experience. To investigate users' intentions to interact with specific referents involves understanding the key role of implementing social interactive features. This is because social presence may lead to different communication outcomes depending on the individual's attitude toward his or her communication partner [2]. While a likable communication partner may increase positive social outcomes, on the contrary, enhancing the social presence of a disliked communication partner could lead to less desirable results [2].

2.2 Product Configuration Environment

The distinctive goal of B2C product customization strategy is to involve customers in the design of the product to meet their individual idiosyncratic needs without a significant increase in production or distribution costs [27] nor substantial trade-offs in quality and time performance [28–31]. Due to the specific characteristics of this strategy, customer decision-making when shopping for a self-designed product is remarkably different from shopping for take-it-or-leave-it products. This is because, at each step of the self-design process, customers have to choose the solution that best matches their needs, and whenever they have no precise knowledge of what solutions might correspond to their needs, choosing among a variety of product solutions can be overwhelming [32]. Paradoxically, product variety results in an excessive amount of information on product configuration solutions which can put users in a condition called choice complexity [4, 33]. When firms attempt to increase their sales by offering more product variety and customization, this may result in loss of sales due to the choice complexity induced by product variety and customization [32]. This is called the product variety paradox.

Information technology plays a critical role in preventing the product variety paradox by better guiding users in their decision-making along the product self-design process via online sales configurators. In particular, knowledge management software such as online sales configurator [34, 35] and recommender systems [36, 37] can profoundly simplify users' tasks by guiding their decision-making and/or suggesting optimal solutions [5, 37, 38]. Online sales configurators (OSCs) are knowledge management software applications that implement mass customization strategies [30, 35] by helping potential customers find an optimal solution. Recommendation systems reduce the risk of product variety paradox because of their ability to reduce choice complexity and proactively support users in their decision-making processes [36, 37, 39, 40] by suggesting complete configurations or ways to complete interim configurations.

Although configurator capabilities and recommender systems can support users by providing a personalized and dynamic dialogue [5, 38, 41], interactions are automatically generated by the system itself (e.g. chat box and recommender algorithms) but in product configurator environment do not enable features that allow human-assisted interactions with different communication partners that users can select whenever they need it.

The purpose of this study is to seek determinants to enrich configurator environments with digital social interactivity and social presence to convey users with additional support to those provided by the configuration capabilities and recommender systems integrated into product configuration environments. To achieve this goal, the study explores a set of users' motivational drivers to interact with a real person to detect which determinants can support their decision-making with social interactive features (e.g. dis/likable communication partners). The relevance of this exploration relies on the boundary conditions of the benefits of increased social presence in terms of interpersonal outcomes of enhanced social presence [2]. As stated by Oh et al. [2], the implementation of social interactivity can benefit user experience, but it can also engender negative responses from socially withdrawn users who may be less motivated to attend to social cues that enhance social presence. While more socially oriented individuals prefer to interact through socially enriched features like audio, video, and face-to-face interactions, less socially oriented individuals may prefer to interact through text-based interactive features [2].

2.3 Customers' Shopping Motivations

Shopping motivations refer to the dispositions of online consumers toward the task of shopping online that are manifested by the expected benefits each consumer seeks to receive from the online store [42].

The literature on customer behaviour describes shoppers as directed by at least three macro areas of shopping motives that drive their decision-making processes: goal-oriented motives [43], experiential-oriented motives [43], and social motives [44]. Individuals shop online differently depending on whether their motivations are primarily experiential (such as enjoying the shopping process and seeking for hedonic or social benefits), goal-oriented (such as looking for product functionalities and functional goals) [45, 46] and/or driven by social motives (such as joining a group, emulating others' behaviours, approving a trend, sharing experiences, and seeking social rewards) [44].

Goal-oriented motivations refer to the utilitarian benefits that customers expect to obtain. For the present study, we focus on convenience search (i.e. better price, product quality, delivery cost, and saving search time) as a key determinant of a customer's effort to choose the product that best suits their cost/benefit criteria [45, 47].

Experientially oriented motivations refer to hedonic benefits that customers expect to obtain. For the present study, we focus on creative stimuli [48]. In the product self-design process, creative stimuli are relevant motivational factors because they are linked to the individual pride of authorship [7]. When self-customizing a product, the individual invests personal effort, time, and attention in defining the characteristics of the product; hence, psychic energy is transferred from the self to the product [49, 50]. In self-designing products, creativity plays a key role in customers' decision-making to create unique products (uniqueness) and products that are representative of those who create them (self-expressiveness) [8].

Social motives refer to the benefits that individuals derive from social interactions defined in literature as the enjoyment of socializing with others as well as shopping with others (e.g. friends, familiar) [51]. Social interactions while shopping also remain a robust motivator of online shopping behaviors [15, 52]. As an example, the influence of friends, family, and colleagues plays a key role both in guiding customers' decision-making processes [53, 54] and in reducing the risk perceived by those who shop online [55].

The present study aims to contribute to the research on customers' behavior in the specific domain of e-commerce for customized products. To study customers' experience when directly engaged in the design of their products is especially relevant. This is because customers may need additional support to their decision-making process by feeling in contact with real persons to achieve the benefits they seek to receive from their configuration/shopping experience.

3 Method

We start this exploration process by considering independently the motivations for interactions with different referents and the interactions at different configuration stages. Given the early stage of research on OSC users' need for social interaction, we engaged in exploratory research to examine users' motivations for interacting with different referents and at different configuration stages.

To analyze the configurator users' motivations for social interaction, we collected 937 configuration experiences made by participants of a sample of 187 potential customers using 378 sales configurators available online.

The collection of configuration experiences was made by assigning a set of five configurators to each participant. Each set was selected based on participants' preferences for specific product types in such a way that each OSC set was different from each participant and can simulate a shopping experience where participants were involved in product configuration. After each experience, a participant filled out a questionnaire.

3.1 Online Sales Configurators Selected for the Study

The sample of 378 online sale configurators was selected from the Cyledge database. This database is the only publicly available list of online sales configurators, and it has been used in previous research on OSCs [25]. Among the 1,252 entries in the database, an initial selection was made according to English as the de facto lingua franca [56] for business [57].

The second step of the selection procedure involved stratified probabilistic sampling. Each stratum was identified by a country–industry–product combination. As an industry-classification list, we used 17 industries that, at the time of the study, were proposed in the database (i.e. Accessories, Apparel, Beauty and Health, Electronics, Food and Packaging, Footwear, Games and Music, House and Garden, Industrial Goods, Kids and Babies, Motor Vehicles, Office and Merchandize, Paper and Books, Pet Supplies, Printing Platforms, Sportswear and Equipment, and Uncategorized).

For each stratum, we randomly chose at least two-thirds of the configurators listed in the database. In the case of fractions, we chose the smallest superior integer. Eventually, the configurators that were no longer active were replaced by active ones, which were randomly chosen from within the same stratum. This procedure recalls the one adopted in a previous study [25].

3.2 Participants to the Study

With the purpose of sampling young adults, we selected management engineering students from the authors' university. Our sample of 187 participants consisted of 129 males and 60 females. The ages of the participants ranged between 22 and 42 years (with an average age of 24 years). Previous research recognized that young people represent the majority of B2C sales configurator users [4].

Before responding to the questionnaire, the participants attended an orientation at a laboratory dedicated to social product configuration systems. There, they were briefed about the meaning and purpose of each statement in the questionnaire. The roles of each referent that participants could choose as a communication partner in case they needed to interact with any real person at each step of the configuration/shopping process via online configurators were also explained.

Any questions or doubts from the participants about the configuration simulation were solved during the orientation laboratory they attended before and while they accomplished the questionnaire. Participants were aware that the shopping process was simulation and that each configurator provided different experiences depending on the product, the specificity of each OSC, and the mass customization capability of each company.

Participants are also profiled as web users to detect their confidence in online shopping. Of the participants, 79.9% had a favorable attitude toward online shopping. In more detail, 47.1% of the participants were web users who made regular purchases on e-commerce websites, 33% were web users who made occasional purchases online (e.g., only in specific product categories), 10.6% were not interested in online shopping, and the remaining 9% did not provide an answer. Each participant filled out a questionnaire after every configuration experience (five per participant).

3.3 Questionnaire

The design of the questionnaire required several tests before drafting the final version. The tests also considered the qualitative feedback provided by a sample of participants interviewed to carry out the pre-test of the questionnaire. To structure the questionnaire, we followed the parallel[1] between the step of configuration/shopping described in Franke et al. [26] and the corresponding step of customer decision-making described in Engel et al. [58]. The uniform formulation of questions (Table 1 column 3) made it possible to graphically design the questionnaire as a table with 27 cells to fill up (Table 1). This way, the participants could fill out the questionnaire without having to reread similar statements/questions several times.

[1] In Engel et al. [57] customers' decision-making process is structured in the following steps: (a) need recognition, (b) alternative evaluation, (c) purchase, and (d) post purchase. Following Franke et al. [25] the configuration process is divided in the following steps: (a) initial idea generation, (b) intermediate evaluation, and (c) final configuration evaluation.

Table 1. Structure of the questionnaire to fill up

Motivational drivers and assigned code	General question to be answered with the following statements: "I felt the need to interact with xxx to…"	Referent types and Configuration Steps								
		"xxx" = My contacts			"xxx" = Experts from the company			"xxx" = Other configurator users		
		Step 1	Step 2	Step 3	Step 1	Step 2	Step 3	Step 1	Step 2	Step 3
Search for Convenience (CONV)	To reach the configuration that best meets my needs and budget									
Creative achievement (CREA)	To get inspirations for my product configuration									
Social feedback (SREW)	To be assured in my configuration choices									

Following parallelism between customers' decision-making process [57]*and product configuration process* [25], *Step 1 refers to the initial product configuration idea, Step 2 refers to the intermediate product configuration (not the definitive one), and Step 3 refers to the final configuration.. Columns 1 and 2 are not present in the questionnaire, however they are reported here to clarify the logical structure of the questionnaire*

The statements refer to users' motivations to digitally interact with three types of referents: (i) individual from users' personal networks (here referred to as "users' contacts" or UXC), (ii) company representatives (here referred to as "experts from the company" or EXC), and (iii) persons unknown to users but with experience in shopping for self-design products (here referred to as "other configurator users" or OCU).

Statements are formulated in a way that users can express their need to interact with the three referent types at each step of their configuration process and evaluate to what extent their need is motivated by the three motivational drivers (Table 1). Each participant was asked to express their level of agreement or disagreement with each proposed statement in the questionnaire using a scale from 1 to 5 (where 1 means completely disagree, 2 disagree, 3 neither agree nor disagree, 4 agree, and 5 completely agree). To avoid the repetition of the three referents in the questionnaire, we graphically refer to each one of the possible referents with this symbol: "xxx" (see Table 1).

At this explorative stage, the study focuses on goal-oriented motivation related to the convenience search to explore to what extent users' motivation to interact with real persons is triggered by the search for the product that best suits the cost/benefit ratio that customers set for themselves [45]. As a result, we formulated the following statement:

- "I felt the need to interact with "xxx" to reach the configuration that best meets my needs and budget."

With experiential-oriented motivations, at this first stage, the study focuses on motivational drivers related to creative achievement to explore to what extent users' motivation to interact with real persons is triggered by their pride to create their own product [50]. As a result, we formulated the following statement:

- "I felt the need to interact with "xxx" to get inspired for my product configuration."

With social motives, at this first stage, the study focuses on motivational drivers related to social feedback to explore to what extent users' motivation to digitally interact with others is triggered by soliciting feedback from real persons. As a result, we formulated the following statement:

- "I felt the need to interact with "xxx" to be assured of my configuration choices."

4 Results

Besides quantitative results, the respondents provided qualitative information by commenting on their answers to the questionnaire on social interaction motivational drivers. The qualitative information was used in this section to interpret the quantitative results.

4.1 Users' Motivations for Digital Social Interaction with Personal Contacts During Product Configuration

Table 2 shows that creative achievement is a motivational driver that triggers users' need to look for social interaction during their self-design process at both steps of initial

Table 2. Users' motivational drivers to interact with their contacts

Users' level of agreement to seek digital interactions with personal contacts (UXC)	Step 1: initial idea development			Step 2: interim Configuration			Step 3: final configuration		
	Motivational drivers			Motivational drivers			Motivational drivers		
	Creative achievement	Convenience search	Reassurance	Creative achievement	Convenience search	Reassurance	Creative achievement	Convenience search	Reassurance
	CREA	CONV	SREW	CREA	CONV	SREW	CREA	CONV	SREW
No answer	0.5%	0.9%	0.5%	0.4%	1.0%	0.4%	0.1%	1.2%	0.1%
Tot. disagree	27.4%	48.5%	39.5%	25.7%	45.7%	34.6%	34.6%	67.3%	24.7%
Disagree	7.8%	17.9%	13.7%	9.4%	17.0%	10.2%	14.3%	15.5%	7.0%
Neutral	16.6%	16.6%	18.1%	21.5%	19.3%	19.9%	22.5%	9.9%	17.2%
Agree	28.3%	12.0%	19.4%	29.6%	12.7%	25.2%	19.9%	4.2%	33.2%
Comp. agree	19.3%	4.2%	8.8%	13.4%	4.4%	9.7%	8.6%	1.9%	17.8%
Tot	100%	100%	100%	100%	100%	100%	100%	100%	100%

Tot.Disagree: totally disagree; Neutral: neither disagree nor agree; Comp.Agree: completely agree

idea development (47.6%) and intermediate configuration (43%). Based on the results, when searching for creative stimuli to inspire them in their product configuration, users' levels of agreement and disagreement to get inspiration from personal contacts are not so different from each other. However, results on users' motives to interact with their contacts are more evident in cases where there are social motives. In 51% of the cases, once the configuration process is close to being finalized (step 3), users seek reassurance from their personal contacts on their decisions on final product configuration.

The need to interact with personal contacts is perceived by participants at each step of the product configuration process to a lower or higher extent depending on the motivational driver and the specific step of product self-design and decision-making. Users' contacts are relied on in a greater degree for motivations concerning social reward and creative stimuli, while, in a much lesser degree, for goal-oriented motivations. In this regard, users clearly express their disagreement on engaged interaction with their contacts for convenience search.

By complementing these results with information derived from interviews, participants expressed that their contacts could advise them both in terms of creative achievement and reassurance in configuration choice before proceeding with the purchase. Conversely, users rarely expect to be advised by their contacts about product convenience budgets and other functional factors. They interact with their contacts more when they need to collect information from trustworthy individuals who are familiar with their personal tastes and habits. The opinions of these users' contacts were also relevant in terms of reassuring users about the esthetic aspects of the configured products.

Some respondents explained that they take into significant consideration the opinions of their contacts because when buying a product, they prefer that the individuals within their circles like it. The respondents also prefer to interact with their contacts prior to making their purchase decisions, as this is when they are interested in being reassured of the suitability of their selected configurations.

4.2 Users' Motivations for Digital Social Interaction with Experts from the Company During Product Configuration

Table 3 reports that the search for convenience in terms of configuration price underlies users' motivation in seeking an expert from the company to an almost equal extent at each step of the product self-design from 36.2% up to 38.7% of cases. To a lesser extent, the number of those who agree and disagree are equal in terms of user's goal achievement.

A limited percentage of users felt the need to interact with company experts for experiential motivations both at initial step 1 (18.9%) and step 2 (16.6%). Being reassured of their configuration choices was a motivational driver only in a few cases (up to 15.9%) at each step of the configuration process. Results show that experiential motivations, such as creative achievement, and social motives such as reward, were not related to users' need to interact with these referents in the majority of the configuration experiences.

By complementing these results with information derived from interviews, participants explained that their desire to interact with company representatives was triggered by their need to gather specific information that only experts from the company could provide. For example, when users need technical information related to the configured

Table 3. Users' motivational drivers to interact with an expert from the company

Users' level of agreement to engage in digital interactions with Experts from the company (EXC)	Step 1: Initial idea development			Step 2: Interim configuration			Step 3: Final configuration		
	Motivational drivers			Motivational drivers			Motivational drivers		
	Creative achievement	Convenience search	Reassurance	Creative achievement	Convenience search	Reassurance	Creative achievement	Convenience search	Reassurance
	CREA	CONV	SREW	CREA	CONV	SREW	CREA	CONV	SREW
No answer	0.6%	1.0%	0.6%	0.5%	0.7%	0.5%	0.2%	0.6%	0.3%
Tot. Disagree	47.8%	37.6%	65.3%	48.8%	33.3%	60.7%	56.9%	35.6%	56.6%
Disagree	15.9%	7.9%	12.8%	15.3%	8.5%	14.1%	18.0%	9.2%	12.7%
Neutral	16.8%	16.5%	12.1%	18.8%	18.7%	13.1%	16.2%	18.4%	14.5%
Agree	13.6%	22.3%	7.2%	13.0%	24.2%	9.0%	6.7%	23.9%	11.2%
Comp. Agree	5.3%	14.7%	2.0%	3.6%	14.5%	2.6%	1.9%	12.3%	4.7%
Tot	100%	100%	100%	100%	100%	100%	100%	100%	100%

Tot.Disagree: totally disagree; Neutral: neither disagree nor agree; Comp.Agree: completely agree

product or the configurator itself, they prefer to interact with a company expert. In addition, users prefer to interact with an expert when they need explanations about the cost or timing of delivery. The need to interact with EXC is motivated by users' need to gather information promptly while they are configuring to enable them to quickly apply changes and continue with the configuration process, especially in the case of high-priced products, such as cars or goods that require a more accurate evaluation by users.

4.3 Users' Motivations for Digital Social Interaction with Other Configurator Users During Product Configuration

Results on users' motivational drivers to interact with other configurator users show that users rely to a lesser extent on the previous two types of referents (Table 4). The users' need to interact with other configurator users is motivated by creative achievement to an equal extent at both the first (28.5%) and second steps (23.4%) of the configuration process. Similar results are registered for the convenience search. In limited cases, users were surprisingly motivated in interacting with OCU for reassurance reasons at the final configuration step (16.9%). With references to the three motivational drivers users have less motivation in interacting with OCU when they have doubts regarding their configuration solutions (step 2) or when they are close to making their final purchase decisions (step 3).

This data is surprising since product self-design environments are mostly connected with communities of users who provide mutual support to each other. Other configurator users are the only available communication partners in addition to the expert from the company reachable via email for customer care services. As a result, research on product configurators mainly focuses on the mutual support found within the community of configurator users. Our results are also in agreement with the conclusions from previous studies on the influence (mainly negative) of the information exchange between users of self-designed products [22]. These first explorative results confirm the key role of recommender systems and configurator capabilities to support those users who may not be interested in interacting with other users.

When complementing results from the questionnaire with information derived from interviews, participants explained that their motivations to interact with other users is related to their need to gather information from a neutral source. The adjective "neutral," as used by respondents, refers to a source that has no interest in pursuing personal advantages, unlike a company representative might. Even so, respondents indicated that they find it difficult to trust the reliability of the comments of someone whom they do not know. The respondents indicated a preference for interacting with other users, for the most part, in cases where they had previous product knowledge. This enables them to compare their knowledge with other users' comments and, thus, assess the reliability of the information provided.

Table 4. Users' motivational drivers to interact with other configurator users

Users' level of agreement to engage digital interactions with other configurator users (OCU)	Step 1: Initial idea development			Step 2: Interim configuration			Step 3: Final configuration		
	Motivational drivers			Motivational drivers			Motivational drivers		
	Creative achievement	Convenience search	Reassurance	Creative achievement	Convenience search	Reassurance	Creative achievement	Convenience search	Reassurance
	CREA	CONV	SREW	CREA	CONV	SREW	CREA	CONV	SREW
No answer	1.5%	2.0%	1.7%	0.5%	1.2%	0.6%	0.2%	1.0%	0.3%
Tot. Disagree	39.1%	49.9%	59.4%	42.5%	50.6%	57.3%	53.1%	50.3%	53.8%
Disagree	12.1%	14.3%	15.9%	14.7%	14.8%	16.8%	15.7%	16.6%	14.9%
Neutral	18.9%	19.2%	14.2%	18.8%	19.7%	15.0%	16.5%	19.4%	14.1%
Agree	20.1%	11.4%	6.9%	18.1%	10.9%	8.3%	11.0%	9.8%	13.7%
Comp. Agree	8.4%	3.1%	1.8%	5.3%	2.8%	1.9%	3.4%	2.9%	3.2%
Tot	100%	100%	100%	100%	100%	100%	100%	100%	100%

Tot.Disagree: totally disagree; Neutral: neither disagree nor agree; Comp.Agree: completely agree

5 Discussions

The present study is one of the first studies on product configurator systems focused on understanding users' need to interact with real people to design user experiences that are enhanced with social presence. It specifically addressed this issue by focusing on users' motivational drivers to interact with one or more communication partners at each step of the product configuration process to ask for support to convenience search, creative achievement, and social reassurance. The study addresses the main research questions: how to integrate social technology into self-design environments to make positive experiences (almost) certain for its users. In responding to the main research question, the study also contributes to the research lines considered in the related work section, as described in the following:

5.1 Social Presence

Since the implementation of social presence leads to different outcomes depending on an individual's attitude towards their communication partner [2], our results contribute to this research line by investigating both the dimensions of "with whom" and at "which step" of the configuration process users seek social interaction with real persons. The results show to what extent three different types of communication partners (personal contacts, experts from the company and other configurator users) become likable or dislikable depending on users' goal-oriented experiential and social motives to interact with real persons at each step of their configuration process. Our results confirm the key role of relevant others (e.g. family, friends, as colleagues) in influencing a user's decision process [59, 60]. Results show that the implementation of social interactive features to enable interaction between users and their relevant others can positively influence user experience, especially whenever these are implemented at the beginning (step 1) and the end of the product configuration process (step 3). At step 1, users seek the social presence of people socially next to them to be supported in their creative achievement, while at step 3, they seek the same kind of communication partners to be reassured on their configuration choice. The results confirm that social information from friends is especially useful for the improvement of recommendation accuracy [60].

The experts from the company are desirable communication partners when, at step 1 of their configuration, users seek convenience in finding solutions that fit with their needs. For the same goal-oriented motivation, but to a lesser extent, EXCs are considered likable partners at steps 2 and 3. During the configuration process, experts from the company results disliked communication partners when users seek creative achievement and social reassurance.

The considered motivations drive only, to a very lesser extent, users in seeking interaction with other configurator users. This third type of communication partner is disliked in most configuration experiences. To a low extent, interactions with these partners are done to seek creative achievement motivation at the first step of product configuration and, to a lesser extent, at the second step.

5.2 Contributions on Customers' Behavior Research Line

The present study contributes to the research on customers' behavior in a technology-mediated environment by exploring these behaviors and their motivational drivers (i.e. convenience search, creative achievement, and reassurance) in the specific domain of eCommerce for customized products. To study customers' experiences when they are directly engaged in the design of their products is especially relevant. This is because customers may need human-assisted support to face the specific decision-making challenges required to self-design a product and thus achieve the benefits they seek to receive from their configuration or shopping experience [42]. This study follows the previous research on human-computer interactions (HCI) that advocate the importance of human-centered design [61] and fulfilling users' non-instrumental needs in providing them with gratifying user experiences. In particular, studies on emotional usability, about '90teens by Logan et al. [62] and more recently by Hassenzahl et al. [63–65] highlighted that HCI must be concerned about aspects of interactive products (i.e. its fit to behavioral goals) as well as about hedonic aspects, such as stimulation (i.e. personal growth, an increase of knowledge and skills), identification (i.e. self-expression, interaction with relevant others).

Accordingly, this explorative study focuses on users' motivational drivers behind their need to interact with real persons in B2C human computer-mediated environments. We found that motivational drivers differ based on "with whom" users have to interact and "at which step" they experience this need to interact. Our findings also highlight the key role of relevant others as desirable communication partners and suggest implementing configurator environments with social interactive features that enable interaction between users and their personal contacts, since social information from people socially next to users (e.g. a friend) proved to be very useful in the improvement of recommendation accuracy [60].

5.3 Contributions to Research Line on Product Configuration Environment

This study contributes to the research line on the B2C product configurator environment. The results of our exploratory research show users' need for human-assisted interactions at each step of their configuration process. Results confirm previous studies on configurator users' need for digital social interaction as experienced in the configuration environment [66]. In addition, results suggest that to maximize the benefits of the implementation of digital social interactive features, it is important for user experience designers to consider this need in terms of "with whom" and "at which step" configurator users experience it. The benefit of implementing social interactivity and social presence on user experience depends on whether or not an individual is socially-oriented. Aside from implementing systemic human-computer interactions into a configurator environment, including socially-interactive features that enable the selection of a desirable partner for human-assisted interaction whenever needed by users can assure a social presence that benefits any type of user. Despite the growing connection between OSC and social software, there is currently no social technology that has been implemented into the product configurator environment to support users in choosing a desirable communication partner for human-assisted interaction whenever they are needed during the configuration process [25].

A recent study that explored configurator users' need for digital interaction with real persons [66] reported that majority of OSC users (88%) experienced the need for social interaction in their configuration experiences. Only 4% of OSC users did not experience a desire to interact with real people in any form during their configuration experiences, while 8% did not provide a definitive answer as to whether or not they perceived this need to be relevant [66]. Moreover, users seek to interact with user contacts (75% of cases), experts from the company (68%), or other configurator users (45%), thus highlighting OSC-user demand for human-assisted consulting during the configuration process [65]. The percentages provided by a recent study [66] indicate that the need to engage in human-assisted interactions varies depending on which type of referent is involved in the interaction (the "with whom" factor). This is unsurprising given that different referents provide different kinds of information and support. However, it raises the question of what determines configurator users' need for social interaction.

The present study moves a step towards elucidating this point by exploring to what extent users' need for digital social interaction relies on the three selected motivational drivers (i.e. convenience search, creative achievement, and social reassurance). The results of the present study show that none of the selected motivational drivers drive this need in more than 50% of users. This suggests that the motivational drivers for social interaction with real people during the configuration process are heterogeneous. Thus, several social interaction features should be provided to cater to different user needs. This complicates the work of online configurator designers.

Finally, the present research has followed an exploratory approach. It aimed to explore the strength of the effect of different motivational drivers in various steps of the configuration process and with other factors. The provided descriptive evidence paves the way for more sophisticated analyses based on inferential statistics. It will be particularly interesting to investigate how the implementation of social presence and interactive features can influence user experience in relation to their digital social interaction needs.

6 Conclusions

Digital transformation and the current health emergency call for a rapid shift from business ecosystems to digital business ecosystems. This transformation also requires companies to be prepared for the challenges of a Web environment where social technologies lead online transactions among users and are influencing their expectations in terms of social presence and digital social interactions.

On one hand, the integration of product configurator systems with social technologies requires companies to acknowledge customers' social interaction needs and implement social technologies accordingly to fulfill their needs during the self-design process. On the other hand, it requires user experience designers to acknowledge what determinants rely on this need to properly provide users with social interactive features that assure (almost) certainly positive experiences for them. The present study adopts a user-centered perspective to seek determinants to enrich configurator environments with digital social interactivity and social presence. These, in turn, support users in engaging human-assisted interactions by choosing among one or more communication partners that can assist them in their search for convenience, creative achievement, and social reward. The results of this study provide vendors with useful suggestions in acknowledging customers' social interaction needs. It also provides user experience designers with insights on how to deliver customer experiences that match customers' actual expectations in terms of social presence. Based on the results of this study, to benefit positive outcomes of social presence enhancement, OSC developers must carefully evaluate determinants such as whom users seek human-assisted interaction, what step they are in their configuration process, and what benefits they aim to achieve from their experience via OSCs.

The results obtained also open the way for strengthening some lines of research on the personalization of users' experience such as (a) the design of digital social interactive features to enable social recommender process relevant to users during product configuration experience and (b) enabling social interactions between configurator users and their relevant others and/or desirable communication partners.

Further research will address the limitations of the present explorative study. The participants in our study constitute a convenience sample and it may be representative only for young adults' potential customers of the considered products. Future research should seek to replicate our findings in truly representative samples of potential customers. Furthermore, each configuration/shopping process was only a simulation and did not end with any effective purchases.

References

1. Matt, C., Hess, T., Benlian, A.: Digital transformation strategies. Bus. Inf. Syst. Eng. **57**(5), 339–343 (2015)
2. Oh, C.S., Bailenson, J.N., Welch, G.F.: A systematic review of social presence: Definition, antecedents, and implications. Front. Rob. AI **5**, 114 (2018)
3. Fitzgerald, M., Kruschwitz, N., Bonnet, D., Welch, M.: Embracing digital technology: a new strategic imperative. MIT Sloan Manag. Rev. **55**(2), 1 (2014)
4. Trentin, A., Perin, E., Forza, C.: Sales configurator capabilities to avoid the product variety paradox: Construct development and validation. Comput. Ind. **64**(4), 436–447 (2013)
5. Felfernig, A., Hotz, L., Bagley, C., Tiihonen, J.: Knowledge-Based Configuration: From Research to Business Cases. Newnes (2014)
6. Mandl, M., Felfernig, A., Teppan, E., Schubert, M.: Consumer decision making in knowledge-based recommendation. J. Intell. Inf. Syst. **37**(1), 1–22 (2011)
7. Trentin, A., Perin, E., Forza, C.: Increasing the consumer-perceived benefits of a mass-customization experience through sales-configurator capabilities. Comput. Ind. **65**(4), 693–705 (2014)
8. Sandrin, E., Trentin, A., Grosso, C., Forza, C.: Enhancing the consumer-perceived benefits of a mass-customized product through its online sales configurator: an empirical examination. Ind. Manag. Data Syst. **117**(6), 1295–1315 (2017)
9. Babin, B.J., Darden, W.R., Griffin, M.: Work and/or fun: measuring hedonic and utilitarian shopping value. J. Cons. Res. **20**(4), 644–656 (1994)
10. Franke, N., Schreier, M.: Why customers value self-designed products: the importance of process effort and enjoyment. J. Prod. Innov. Manag. **27**(7), 1020–1031 (2010)
11. Franke, N., Schreier, M., Kaiser, U.: The "I designed it myself" effect in mass customization. Manag. Sci. **56**(1), 125–140 (2010)
12. Kamis, A., Koufaris, M., Stern, T.: Using an attribute-based decision support system for user-customized products online: an experimental investigation. MIS Q. **32**, 159–177 (2008)

13. Jones, M.A., Reynolds, K.E., Arnold, M.J.: Hedonic and utilitarian shopping value: investigating differential effects on retail outcomes. J. Bus. Res. **59**(9), 974–981 (2006)
14. Gruber, T.: Collective knowledge systems: where the social web meets the semantic web. Web Semant. Sci. Serv. Agents World Wide Web **6**(1), 4–13 (2008)
15. Huang, Z., Benyoucef, M.: User preferences of social features on social commerce websites: an empirical study. Technol. Forecast. Soc. Change **95**, 57–72 (2015)
16. Biocca, F., Levy, M.R.: Communication in the Age of Virtual Reality. Routledge, Abingdon (2013)
17. Gefen, D., Straub, D.: Managing user trust in B2C e-services. e-Service **2**(2), 7–24 (2003)
18. Lu, B., Fan, W., Zhou, M.: Social presence, trust, and social commerce purchase intention: an empirical research. Comput. Hum. Behav. **56**, 225–237 (2016)
19. Kumar, N., Benbasat, I.: Shopping as experience and website as a social actor: web interface design and para-social presence. In: ICIS 2001 Proceedings, vol. 54 (2001)
20. Hassanein, K., Head, M.: Manipulating perceived social presence through the web interface and its impact on attitude towards online shopping. Int. J. Hum Comput Stud. **65**(8), 689–708 (2007)
21. Jeppesen, L.B.: User toolkits for innovation: consumers support each other. J. Prod. Innov. Manag. **22**(4), 347–362 (2005)
22. Moreau, C.P., Herd, K.B.: To each his own? How comparisons with others influence consumers' evaluations of their self-designed products. J. Cons. Res. **36**(5), 806–819 (2009)
23. Hildebrand, C., Häubl, G., Herrmann, A., Landwehr, J.R.: When social media can be bad for you: community feedback stifles consumer creativity and reduces satisfaction with self-designed products. Inf. Syst. Res. **24**(1), 14–29 (2013)
24. Schlager, T., Hildebrand, C., Häubl, G., Franke, N., Herrmann, A.: Social product-customization systems: Peer input, conformity, and consumers' evaluation of customized products. J. Manag. Inf. Syst. **35**(1), 319–349 (2018)
25. Grosso, C., Forza, C., Trentin, A.: Supporting the social dimension of shopping for personalized products through online sales configurators. J. Intell. Inf. Syst. **49**(1), 9–35 (2017)
26. Franke, N., Keinz, P., Schreier, M.: Complementing mass customization toolkits with user communities: how peer input improves customer self-design. J. Prod. Innov. Manag. **25**(6), 546–559 (2008)
27. McCarthy, I.P.: Special issue editorial: the what, why and how of mass customization. Prod. Plan. Control **15**(4), 347–351 (2004)
28. Pine, B.J.: Mass Customization: The New Frontier in Business Competition. Harvard Business Press, Boston (1993)
29. Liu, G., Shah, R., Schroeder, R.G.: Linking work design to mass customization: a sociotechnical systems perspective. Decis. Sci. **37**(4), 519–545 (2006)
30. Trentin, A., Perin, E., Forza, C.: Product configurator impact on product quality. Int. J. Prod. Econ. **135**(2), 850–859 (2012)
31. Trentin, A., Perin, E., Forza, C.: Overcoming the customization-responsiveness squeeze by using product configurators: beyond anecdotal evidence. Comput. Ind. **62**(3), 260–268 (2011)
32. Forza, C., Salvador, F.: Application support to product variety management. Int. J. Prod. Res. **46**(3), 817–836 (2008)
33. Valenzuela, A., Dhar, R., Zettelmeyer, F.: Contingent response to self-customization procedures: implications for decision satisfaction and choice. J. Mark. Res. **46**(6), 754–763 (2009)
34. Felfernig, A.: Standardized configuration knowledge representations as technological foundation for mass customization. IEEE Trans. Eng. Manag. **54**(1), 41–56 (2007)
35. Salvador, F., Forza, C.: Principles for efficient and effective sales configuration design. Int. J. Mass Customisation **2**(1–2), 114–127 (2007)

36. Falkner, A., Felfernig, A., Haag, A.: Recommendation technologies for configurable products. AI Mag. **32**(3), 99–108 (2011)
37. Tiihonen, J., Felfernig, A.: Towards recommending configurable offerings. Int. J. Mass Customisation **3**(4), 389–406 (2010)
38. Tiihonen, J., Felfernig, A.: An introduction to personalization and mass customization. J. Intell. Inf. Syst. **49**(1), 1–7 (2017)
39. Jameson, A., Willemsen, M.C., Felfernig, A., de Gemmis, M., Lops, P., Semeraro, G., Chen, L.: Human decision making and recommender systems. In: Recommender Systems Handbook, pp. 611–648. Springer, Heidelberg (2015)
40. Felfernig, A., Teppan, E., Gula, B.: Knowledge-based recommender technologies for marketing and sales. Int. J. Pattern Recogn. Artif. Intell. **21**(2), 333–354 (2007)
41. Ardissono, L., Felfernig, A., Friedrich, G., Goy, A., Jannach, D., Petrone, G., Schafer, R., Zanker, M.: A framework for the development of personalized, distributed web-based configuration systems. AI Mag. **24**(3), 93–108 (2003)
42. Pappas, I.O., Kourouthanassis, P.E., Giannakos, M.N., Lekakos, G.: The interplay of online shopping motivations and experiential factors on personalized e-commerce: a complexity theory approach. Telematics Inform. **34**(5), 730–742 (2017)
43. Bridges, E., Florsheim, R.: Hedonic and utilitarian shopping goals: the online experience. J. Bus. Res. **61**(4), 309–314 (2008)
44. Solomon, M.R., Dahl, D.W., White, K., Zaichkowsky, J.L., Polegato, R.: Consumer Behavior: Buying, Having and Being. Pearson, London (2014)
45. Rohm, A.J., Swaminathan, V.: A typology of online shoppers based on shopping motivations. J. Bus. Res. **57**(7), 748–757 (2004)
46. Dholakia, U.M., Kahn, B.E., Reeves, R., Rindfleisch, A., Stewart, D., Taylor, E.: Consumer behavior in a multichannel, multimedia retailing environment. J. Interact. Mark. **24**(2), 86–95 (2010)
47. Novak, T.P., Hoffman, D.L., Yung, Y.-F.: Measuring the customer experience in online environments: a structural modeling approach. Mark. Sci. **19**(1), 22–42 (2000)
48. Varma Citrin, A., Sprott, D.E., Silverman, S.N., Stem, D.E., Jr.: Adoption of internet shopping: the role of consumer innovativeness. Ind. Manag. Data Syst. **100**(7), 294–300 (2000)
49. Belk, R.W.: Possessions and the extended self. J. Consum. Res. **15**(2), 139–168 (1988)
50. Schreier, M.: The value increment of mass-customized products: an empirical assessment. J. Consum. Behav. **5**(4), 317–327 (2006)
51. Arnold, M.J., Reynolds, K.E.: Hedonic shopping motivations. J. Retail. **79**(2), 77–95 (2003)
52. Lueg, J.E., Finney, R.Z.: Interpersonal communication in the consumer socialization process: scale development and validation. J. Mark. Theory Pract. **15**(1), 25–39 (2007)
53. Childers, T.L., Rao, A.R.: The influence of familial and peer-based reference groups on consumer decisions. J. Consum. Res. **19**(2), 198–211 (1992)
54. Wang, X., Yu, C., Wei, Y.: Social media peer communication and impacts on purchase intentions: a consumer socialisation framework. J. Interact. Mark. **26**(4), 198–208 (2012)
55. Pires, G., Stanton, J., Eckford, A.: Influences on the perceived risk of purchasing online. J. Consum. Behav. **4**(2), 118–131 (2004)
56. Jenkins, J.: English as a lingua franca: interpretations and attitudes. World Englishes **28**(2), 200–207 (2009)
57. De Swaan, A.: Words of the World: The Global Language System. John Wiley & Sons, Hoboken (2013)
58. Engel, J.F., Blackwell, R., Miniard, P.: Customer Behavior. Dryden, Hinsdale (1990)
59. Chen, A., Lu, Y., Wang, B.: Customers' purchase decision-making process in social commerce: a social learning perspective. Int. J. Inf. Manag. **37**(6), 627–638 (2017)
60. Tang, J., Hu, X., Liu, H.: Social recommendation: a review. Soc. Netw. Anal. Min. **3**(4), 1113–1133 (2013)

61. Leitner, G.: Why is it called human computer interaction, but focused on computers instead? In: The Future Home is Wise, Not Smart, pp. 13–24. Springer, Heidelberg (2015)
62. Logan, R.J., Augaitis, S., Renk, T.: Design of simplified television remote controls: a case for behavioral and emotional usability. In: Proceedings of the Human Factors and Ergonomics Society Annual Meeting, pp. 365–369. SAGE Publications, Los Angeles (1994)
63. Hassenzahl, M., Tractinsky, N.: User experience-a research agenda. Behav. Inf. Technol. **25**(2), 91–97 (2006)
64. Hassenzahl, M., Beu, A., Burmester, M.: Engineering joy. IEEE Softw. **18**(1), 70–76 (2001)
65. Hassenzahl, M.: The thing and I: understanding the relationship between user and product. In: Funology, vol. 2, pp. 301–313. Springer, Heidelberg (2018)
66. Grosso, C., Forza, C.: Users' social-interaction needs while shopping via online sales configurators. Int. J. Ind. Eng. Manag. **10**(2), 139–154 (2019)

Impact of the Application of Artificial Intelligence Technologies in a Content Management System of a Media

Ignacio Romero[1], Jorge Estrada[2], Angel L. Garrido[1](✉), and Eduardo Mena[3]

[1] Henneo Corporación Editorial, Zaragoza, Spain
{ifromero,algarrido}@henneo.com
[2] Hiberus Tecnologías Diferenciales, S. L., Zaragoza, Spain
jestrada@hiberus.com
[3] University of Zaragoza, Zaragoza, Spain
emena@unizar.es

Abstract. Nowadays, traditional media are experiencing a strong change. The collapse of advertising-based revenues on paper newspapers has forced publishers to concentrate efforts on optimizing the results of online newspapers published on the Web by improving content management systems. Moreover, if we put the focus on small or medium-sized media, we find the additional problem of the shortage of single users, very necessary to properly model recommendation systems that help increase the number of visits and advertising impacts. In this work, we present an approach for performing automatic recommendation form news in this hard context combining matrix factoring and semantic techniques. We have implemented our solution in a modular architecture design to give flexibility to the creation of elements that take advantage of these recommendations, and also with great monitoring possibilities. Experimental results in real environments are promising, improving outcomes regarding traffic redirection and clicks on ads.

1 Introduction

In recent years, in practically any media in the world, there are a number of problems that place these types of companies in a difficult situation. The significant decrease of advertising in printed editions and radical changes in the way that readers and profits are achieved have forced traditional media to transform and adapt to a different type of business, with new actors and new rules [1].

The decrease in revenues, derived from the decline in sales of printed newspapers, forces to look for new forms of income in a digital world, where advertising is highly segmented and even personalized for each reader. Although there are many content management systems (CMS) for news in the market, many times these kind of software products does not have enough power to optimize the management of news and advertising needed today [2].

The purpose of this work is to analyze and to study the influence of Artificial Intelligence (AI) technologies on a content management system of a small or

M. Stettinger et al. (Eds.): ISMIS 2020, SCI 949, pp. 139–152, 2021.
https://doi.org/10.1007/978-3-030-67148-8_11

medium size media, and check to what extent it can improve efficiency and the effectiveness of the processes.

To carry out this investigation in a rigorous way, on the one hand, we propose the implementation of a specialized system capable of using AI technologies, and that will allow a fine tuning with the goal of knowing which factors are the ones that most influence the achievement of good results. On the other hand, given the specific nature of this study, a real data and a real scenery will be required. This is an usually complicated aspect, since in many cases it is difficult to access private systems for the realization of experiments. To overcome this difficulty, this work has been carried out jointly by the research and development team of *Henneo Corporacion Editorial*, a company part of HENNEO,[1] a well-known Spanish media group. Thanks to the participation of a company in this research work, the experiments can be conducted on real data, and they can be applied on a completely actual CMS software currently used by some of leading Spanish media.

To that end, it is proposed to design an architecture and to develop a series of AI subsystems, which, integrated with a CMS, are able to provide a series of services that contribute in several aspects:

1. Gaining knowledge of the digital newspaper user's habits.
2. Improving the user experience by adapting the contents to the user's interests.
3. Creating new advertising channels, replacing traditional media.
4. Enhancing results for publishing units by the creation of more segmented and personalized advertising.

Therefore, this approach has the purpose of closing the cycle between content generation and the observation of reader behaviors, integrating AI technologies to automate tasks and personalize content. Besides, it contributes to the transformation of the editorial units towards the new technological era.

This paper is structured as follows. Section 2 analyzes and describes the state of the art. Section 3 explains the architecture proposed for the design of an intelligent CMS. Section 4 explains the AI methods used by the system. Section 5 show and discusses the preliminary results of our tests with real data. Finally, Sect. 6 provides our conclusions and future work.

2 State of the Art

Since the design and definition of the firsts CMS [3], its use in different sectors has been extended, and its use to design news websites is especially valuable [4]. In this context, for the recommendation of content to the readers of the digital media, different approaches have been studied [5]:

1. *Use of common descriptors for describing users and news:* For example, using the section of the media where the piece of news appears (sports, music,

[1] https://www.henneo.com/.

economy, etc.), or using a list of topics covered by the piece of news and for which a user is interested. In this regard, the approximations are based on statistical techniques, natural language processing, and semantics [6].

2. *Automatic classification methods*: They are made through a combination of machine/deep learning and natural language processing methods to assign one or more topics to a piece of news [7].
3. *Segmentation of users, and news classification using this set of segments:* It can be done directly, for example, users interested in economics (that they could be those who have seen more than three economic news in the last month). As an alternative to the creation of user groups manually, clustering techniques can be used, where the users of each group have more similar properties among themselves than with the other users [8]. The disadvantage is to have to decide in advance the number of clusters to be generated, and mainly that the descriptors of the groups thus generated tend to have a dispersion of user interest among the majority of the sections, thereby losing the discriminatory value. In addition, the groups generated are often not homogeneous and show great dispersion among the users of the group. Another methods such as k-means do not allow a user to belong to more than one group.
4. *Statistical methods for automatically assigning topics to news:* For example, using Latent Dirichlet Allocation (LDA) models. LDA has the advantage of not needing to designate possible topics beforehand, because the model generates statistically defined topics by the probability of finding certain words in the news, and is fairly used for hybrid recommenders [9].

Two classic, but effective approaches, are collaborative filtering (CF) and content-based (CB) [10]. Both approaches work well, but the first requires a large number of unique users, which is generally a problem in small media. Besides, users of news platforms do not usually rate the news directly. The CB filter also has limitations since it is necessary to model users very well, which is often complex. News must be also automatically tagged, since today there are no time or resources to do it on a manual basis.

Finally, both methods suffer from the so-called "cold-start problem", especially CF. The cold-start problem happens when it is not possible to make useful recommendations because of an initial lack of ratings. These problems can in turn be differentiated into three typologies: new item, new community, and new user. The last kind has been always the most important regarding actual recommender systems.

Since new users have not yet provided any rating, they cannot receive any personalized recommendations based on a memory-based CF. As soon as the users introduce their firsts ratings they expect the recommender system to offer them personalized recommendations. But it is not feasible because the number of ratings entered is usually not yet sufficient to produce trustworthy recommendations. Therefore, some strategy to alleviate this problem is always necessary [11]. The use of techniques based on Natural Language Processing and Semantics has been seen as an important tool to improve this type of deficiencies, finding a large number of systems that use these technologies [12–14].

3 Architecture

This section describes the proposed architecture with the purpose of allowing the achievement of the objectives listed in Sect. 1. As shown in Fig. 1, the proposed architecture, called "Prometeo", consists of seven main components, integrated into the media CMS:

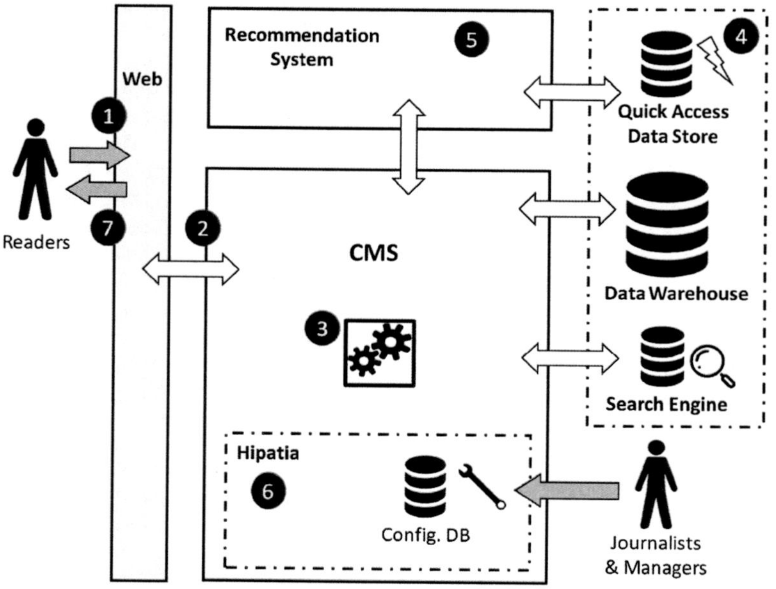

Fig. 1. System architecture proposed, with the seven main components: Acquisition Layer (1), Message queue (2), Processing Unit (3), Storage Units (4), Contents Recommendation Module (5), Management Layer (6), and Output Layer (7).

1. *Acquisition Layer:* The first element is the entry point of the streaming system. It is in charge of recovering information from the external users ("readers") of the newspaper's web produced by the CMS System. Through this entry it is possible to recover very valuable data associated with the user at the moment in which he/she interacts with the media website.
2. *Message queue:* All data sent from the acquisition layer is managed and synchronized by this component. Sending the events to this queue assures the delivery of the data and guarantees persistence.
3. *Processing Unit:* The real-time processing engine in charge of dealing with the different events that occur and managing the information saved in the Storage Units.

4. *Storage Units:* The data processed by the processing layer are saved in different data storage systems, depending on the capacity and speed required. The three types used are detailed below:
   - *Data Warehouse:* It is where the raw data is ingested in order to be exploited by performing ETL[2] tasks and analytics. The contents are extracted from it by performing an ETL to obtain the contents.
   - *Quick Access Data Store:* It is a document database whose purpose is to store the information of contents to show to the readers. The recommended content is composed of the articles of the sites, and this content is extracted from the Data Warehouse by performing an ETL.
   - *Search Engine:* Aggregated data are stored extracted from the Data Warehouse to be consumed by Hipatia. These data are the statistics of the widgets[3] such as pages on which they are displayed, sites associated with the widget, user's events (clicks, impressions, or page shows), images, links, etc.
5. *Contents Recommendation Module:* This sub-system component will periodically assign a list of the sections or articles with the highest propensity expected for each user.
6. *Management Layer:* This component, called *Hipatia*, is used by internal users (managers of the system, or journalists in the media) on the one hand to display in a proper way the data that have been processed, and on the other hand to configure the widgets that shows recommended specific content to the external readers.
7. *Output Layer:* The last component is responsible for displaying the information that the systems have selected as interesting to the readers, by using the widgets defined in the Management Layer.

4 Artificial Intelligence Methods

The recommendation system has been made by designing a hybrid system that uses both CF and CB filtering. On the one hand, CF recommend news based on what similar readers have read. News assessment by the readers is not functional, so news readings are taken into account as the rating. If two readers have similar preferences, news that the first reader reads might interest the second. On the other hand, CB filter suggests news based on a description of the profile of the user's preference and the news description. The CB approach can alleviate the cold-start problem with new users, whose reading history is limited or non existent.

[2] ETL (Extract, transform and load) is the process that allows organizations to move data from multiple sources, process, and load them into another data storage to analyze, or in another operating system to support a business process.

[3] A widget typically is a relatively simple and easy-to-use software application or component made for one or more different software platforms. A web widget is a portable application to offer site visitors shopping, advertisements, videos, or other simple functionality from a third party publisher.

One of the most popular algorithms to solve clustering problems (and specifically for CF approaches) is called Matrix Factorization (MF), a way of taking a sparse matrix of users and ratings, and factoring out a lower-rank representation of both. In its simplest form, it assumes a matrix $A \in A^{m \times n}$ of ratings given by m users to n items. As can be seen in Fig. 2, applying this technique on A will end up factorizing A into two matrices $X \in R^{m \times k}$ and $Y \in R^{n \times k}$ such that $A \approx X \times Y$. Alternating Least Square (ALS) is a MF algorithm built for large-scale CF problems [15]. ALS works very well to solve scalability and sparseness of the ratings data, and it's simple and scales properly to very large datasets.

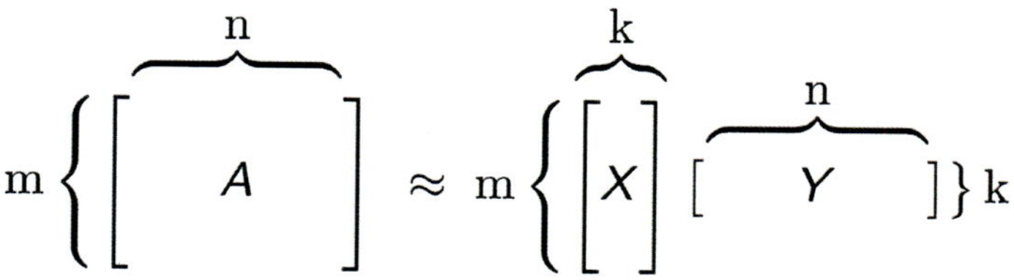

Fig. 2. Matrix Factorization: given a matrix A with m rows and n columns, his factorization is a decomposition of A into matrices X and Y. X has the same number of rows as A, but only k columns. The matrix Y has k rows and n columns, where k is equal to the total dimension of the embedded features in A.

The proposed recommendation system adapts the Weighted Alternating Least Squares (WALS) [16], a weighted version of ALS that uses a weight vector which can be linearly or exponentially scaled to normalize row and/or column frequencies. As in news sites there aren't any sort of user-rated items, it's been decided to rate '1' if the user has visited a piece of news and '0' otherwise. But it's uncertain if a 0 rating means that the reader doesn't like an article or doesn't know about it. Moreover, in many media websites users have few recurring visits, so it is very likely that the number of read news is very low, which makes the recommendation process even more difficult.

In order to palliate these problems, we have improved the CF algorithm with a CB approach, a strategy that provides good results in different contexts [17–19] . On the one hand, those news that are similar to what the user has already read, have been also scored according to its degree of similarity with those read. To assess their resemblance, a similarity metric between them has been used based on the *entities* of each piece of news, by receiving a weighted score between 0 and 1, where zero means no match, and 1 stands for the news are completely equivalent.

The process of obtaining entities, disambiguating them, and assessing the similarity between news has been carried out through an adaptation of NEREA [20], whose general idea is to perform a semantic named entity recognition [21] and disambiguation tasks by using three types of knowledge bases:

local classification resources, global databases (like DBpedia), and its own catalog. The methodology of NEREA has been experimentally tested in real environments and has been successfully applied for example to improve the quality of automatic infobox generation tasks [22].

The similarity between news is evaluated using Bags of Words (BOW) generated thorough the most relevant words located in the same sentences to the named entities of the piece of news. Then, a context vector ($V_{context}$) is built with weights previously calculated by classical TF-IDF algorithm, and this vector is compared with the candidate vectors ($V_{candidate}$) using the cosine similarity, a common method used to measure the similarity using the BOW model:

$$sim(V_{context}, V_{candidate}) = \cos\theta = \frac{V_{context} \cdot V_{candidate}}{|V_{context}||V_{candidate}|}$$

$$= \frac{\sum_{i=1}^{n} V_{context}(i) V_{candidate}(i)}{\sqrt{\sum_{i=1}^{n} V_{context}(i)^2}\sqrt{\sum_{i=1}^{n} V_{candidate}(i)^2}}$$

This method provides a value between 0 and 1 that will be used to assign an "artificial" rating to the rest of the news depending on the degree of similarity with the news that has been read. If more than one piece of news has been read, the process is repeated for each of them, choosing the highest rating if there is more than one artificial rating above zero for the same unread piece of news.

Furthermore, the system directly recommends the most read news to new users, who have not read anything, and thus, there is no information to offer a consistent recommendation. With this double approach, the well-known problem of cold-start is alleviated.

5 Testing

The experiments, as mentioned in Sect. 1, have been carried out in collaboration with HENNEO,[4] one of the main publishing groups in Spain. It is the seventh Spanish communication group by turnover and one of the main audience groups in its category. It also stands out for its continuous collaborations with research, especially on issues of Artificial Intelligence related to Natural Language Processing and Semantics [23–26]. The following describes how the tests on the proposed architecture have been carried out, from its implementation to its evaluation.

5.1 Tools

The implementation of the system has been performed on Xalok,[5] the CMS which manages the media websites of HENNEO using the following tools, most of them belonging to the Google Cloud suite:

[4] https://www.henneo.com/.
[5] https://www.xalok.com/.

- DataFlow[6]: It is about a fully managed service for executing Apache Beam pipelines within the Google Cloud Platform ecosystem. We have used to implement the Processing Unit.
- Cloud DataStore[7]: It is a highly scalable NoSQL database. It acts as the Quick Access Data Store.
- BigQuery[8]: It is a Restful web service that enables interactive analysis of massive datasets, working in conjunction with Google Storage. It has been used as the Data Warehouse element of the proposed architecture.
- Elastic Search[9]: It is a distributed, open source search and analytics engine for any type of data. It acts as the Search Engine of the proposed system.
- AppEngine[10]: Application server where the Hipatia front (Management Layer) is hosted. It has also been used for implementing the Acquisition Layer.
- Firestore[11]: It is in charge of Hipatia application data.

As can be seen in Fig. 3, for the implementation a hybrid architecture (Lamda/Kappa) [27] was chosen. The recommendation system has been adapted from the Weighted Alternating Least Squares (WALS) MF algorithm implemented in Tensorflow[12].

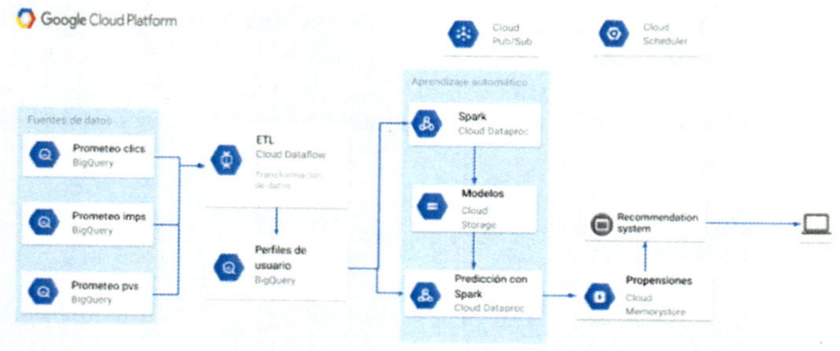

Fig. 3. In the implementation of the system, a hybrid architecture (LAMBDA + KAPPA) was chosen since there was on the one hand a speed layer of data to Bigquery, and on the other a continuous Batch with Dataflow to update the DataStore.

Regarding results presentation, Fig. 4 shows an example of the dashboards that have been developed. These dashboards will include the main metrics that

[6] https://cloud.google.com/dataflow/.
[7] https://cloud.google.com/datastore/.
[8] https://cloud.google.com/bigquery/.
[9] https://www.elastic.co/.
[10] https://cloud.google.com/appengine/.
[11] https://firebase.google.com/products/firestore/.
[12] https://www.tensorflow.org.

indicate the overall performance of the page such as: users, sessions and page views, as well as conversion indicators such as: pages/session, % bounces, and session duration.

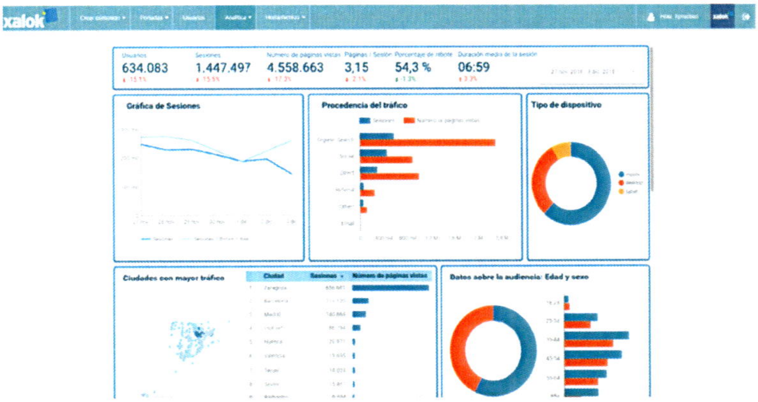

Fig. 4. Example of a dashboard. The data can be analyzed from different dimensions such as: devices, source or source of access of the sessions, time of access, and urls to which users access.

5.2 Dataset and Metrics

The dataset used in the experiments contains real news from the CMS during a period of 7 days. To get data with enough information it's been filtered with two conditions: users who have read at least 5 news, and news read by more than 100 users, resulting in more 300K users 2K news. For privacy reasons the dataset cannot be made public, but it can be shared with other researchers through a collaboration agreement with the company. The metrics used to evaluate the algorithm are RMSE (Root Mean Square Error) and recall. Recall only considers the positively rated articles within the top M, where a high recall with lower M will be a better system.

5.3 Recommendation System Experiments

In Fig. 5 we can see how the system produces personalized recommendations for users. Below we will describe the experimental results with real users and websites, derived from tests carried out at HENNEO.

As seen in Fig. 6 the initial test calculates the recall for four different situations: using a random choice of articles, the most viewed articles, WALS model, and the complete approach. Results obtained using the test dataset shows an

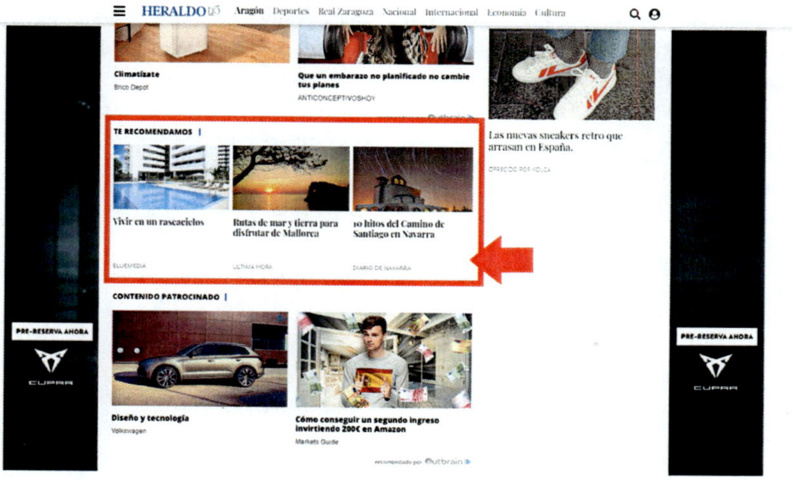

Fig. 5. At the foot of the news published on the HENNEO website www.heraldo.es, the recommendations generated by the Prometeo system appear. In the image they have been marked with a box.

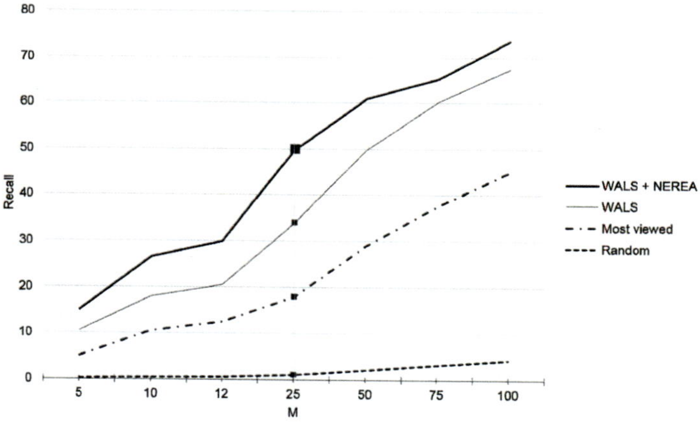

Fig. 6. Results of the recommendation system experiments. Both WALS and the combination of WALS with NEREA improves the results of other approaches.

improvement in recall for a fixed number of recommendations (M) of 25 between the most viewed articles (17.89%), the WALS algorithm (33.9%), and the WALS algorithm improved with NEREA (47.3%).

5.4 Global Results

The improvements and changes that are being made in the system, as well as their effects on the different advertising campaigns that were carried out through this last year have been monitored to evaluate their effectiveness over several websites of HENNEO: *20 minutos*,[13] *Heraldo de Aragón*,[14] and *La Información*,[15] all of them well known Spanish media.

Table 1 shows how results, in this case advertising, have improved almost a 138% throughout the year 2019 after the implementation of the Prometeo recommendation system. Regarding the results of the transfer of user between websites ("bounce") through links generated by Prometeo, an example of the results obtained between two of these digital media of HENNEO is shown in Table 2.

Table 1. Cumulative results by sector of the advertising campaigns recommended in 2019 through the Prometeo's recommendation system. The second column indicates the total number of times that the ad has been accessed through a recommendation link, and the third column shows the percentage of total hits, i.e. how many of the total of impacts have been made thanks to the recommendation system.

Sector	RS Ad hits	% RS
Concerts and shows	17,793	42.43%
Foods & Drinks	37,574	92.54%
Financial	2,047	51.26%
Pay-per-view television	19,639	32.29%
Pharmacy	43,680	87.97%
Real state	7,040	22.39%
Sports	26,783	77.95%
Supply companies	16,532	63.74%
Technology	17,978	66.10%
Telecom	14,983	45.80%
TOTAL	207,072	57.83%

Table 2. Results of the transfer of users between two websites ("bounce") through recommendations during 2019.

Site	Sessions	Unique users
20minutos.es	411,682	105,302
Heraldo de Aragón	1,671,421	696,416

[13] https://www.20minutos.es/.

[14] https://www.heraldo.es/.

[15] https://www.lainformacion.com/.

6 Conclusions and Future Work

In this work, we have studied and quantified the effects of some AI techniques applied on a CMS dedicated to the publication of news in a medium size media. The lack of ratings, and the scarceness of unique and repetitive users, hinders the application of known techniques. Therefore, it is necessary to look for some novel solution that, through a efficient algorithm, can provide good results in the tasks of recommendation, so necessary to increase the number of visits and the impacts on the advertisements.

The main contribution of this work is the design of a modular architecture that can be applied on any news CMS. On the one hand, it allows to integrate a widget generation system, and on the other hand, it adds a recommendation system that works in sync with the rest of the architectural elements.

In addition, Prometeo, the proposed recommendation system improves the possibilities of the known approaches to alleviate the lack of data in these types of environments thanks to the artificial generation of ratings through a system of semantic recognition of the entities described in the news, so it avoids errors due to ambiguous language.

The proposed architecture has the advantage of allowing to incorporate new methods to enhance the system management with minimal effort, and besides it is a language independent platform. The first tests performed over real data and real websites show very good results.

There are many future lines to explore, so the system can achieve a better performance: For example the recommendation could probably be improved if it used more information to set the article rates, such as consideration of temporary factors in the news. The future incorporation of digital subscriptions will also allow to know the readers of the news much better, being able to introduce new techniques that further improve the results of the recommendation.

Acknowledgments. This research work has been supported by the project "CMS Avanzado orientado al mundo editorial, basado en técnicas big data e inteligencia artificial" (IDI-20180731) from CDTI Spain. and the CICYT TIN2016-78011-C4-3-R (AEI/FEDER, UE). We want also to thank all the team of Henneo Corporación Editorial for their collaboration in this work.

References

1. Angelucci, C., Cagé, J.: Newspapers in times of low advertising revenues. Am. Econ. J. Microecon. **11**(3), 319–364 (2019)
2. Zhang, S., Lee, S., Hovsepian, K., Morgia, H., Lawrence, K., Lawrence, N., Hingle, A.: Best practices of news and media web design: an analysis of content structure, multimedia, social sharing, and advertising placements. Int. J. Bus. Anal. **5**(4), 43–60 (2018)
3. Han, Y.: Digital content management: the search for a content management system. Libr. Hi Tech **22**(4), 355–365 (2004)
4. Benevolo, C., Negri, S.: Evaluation of content management systems. Electron. J. Inf. Syst. Eval. **10**(1) (2007)

5. Karimi, M., Jannach, D., Jugovac, M.: News recommender systems–survey and roads ahead. Inf. Process. Manag. **54**(6), 1203–1227 (2018)
6. Altınel, B., Ganiz, M.C.: Semantic text classification: a survey of past and recent advances. Inf. Process. Manag. **54**, 1129–1153 (2018)
7. Zhang, S., Yao, L., Sun, A., Tay, Y.: Deep learning based recommender system: a survey and new perspective. ACM Comput. Surv. (CSUR) **52**(1), 1–38 (2019)
8. Li, Q., Kim, B.M.: Clustering approach for hybrid recommender system. In: Proceedings of the IEEE/WIC International Conference on Web Intelligence, pp. 33–38. IEEE (2003)
9. Chang, T.M., Hsiao, W.F.: LDA-based personalized document recommendation. In: PACIS (2013)
10. Bobadilla, J., Ortega, F., Hernando, A., Gutiérrez, A.: Recommender systems survey. Knowl.-Based Syst. **46**, 109–132 (2013)
11. Gope, J., Jain, S.K.: A survey on solving cold start problem in recommender systems. In: Proceedings of the International Conference on Computing, Communication and Automation, pp. 133–138. IEEE (2017)
12. Albanese, M., d'Acierno, A., Moscato, V., Persia, F., Picariello, A.: A multimedia semantic recommender system for cultural heritage applications. In: Proceedings of the International Conference on Semantic Computing, pp. 403–410. IEEE (2011)
13. Garrido, A.L., Pera, M.S., Ilarri, S.: SOLE-R: a semantic and linguistic approach for book recommendations. In: Proceedings of the 14th International Conference on Advanced Learning Technologies, pp. 524–528. IEEE (2014)
14. Amato, F., Moscato, V., Picariello, A., Piccialli, F.: SOS: a multimedia recommender system for online social networks. Fut. Gener. Comput. Syst. **93**, 914–923 (2019)
15. Zhou, Y., Wilkinson, D., Schreiber, R., Pan, R.: Large-scale parallel collaborative filtering for the netflix prize. In: Proceedings of the International Conference on Algorithmic Applications in Management, pp. 337–348. Springer, Heidelberg (2008)
16. Pan, R., Zhou, Y., Cao, B., Liu, N., Lukose, R., Scholz, M., Yang, Q.: One-class collaborative filtering. In: Proceedings of the IEEE/WIC International Conference on Data Mining, pp. 502–511. IEEE (2008)
17. Hu, R., Pu, P.: Enhancing collaborative filtering systems with personality information. In: Proceedings of the ACM Conference on Recommender systems, pp. 197–204. ACM (2011)
18. Fernández-Tobías, I., Braunhofer, M., Elahi, M., Ricci, F., Cantador, I.: Alleviating the new user problem in collaborative filtering by exploiting personality information. User Model. User-Adapt. Interact. **26**(2–3), 221–255 (2016)
19. Yang, S., Korayem, M., AlJadda, K., Grainger, T., Natarajan, S.: Combining content-based and collaborative filtering for job recommendation system: a cost-sensitive statistical relational learning approach. Knowl.-Based Syst. **136**, 37–45 (2017)
20. Garrido, A.L., Ilarri, S., Sangiao, S., Gañán, A., Bean, A., Cardiel, O.: NEREA: named entity recognition and disambiguation exploiting local document repositories. In: Proceedings of the IEEE International Conference on Tools with Artificial Intelligence, pp. 1035–1042. IEEE (2016)
21. Sekine, S., Ranchhod, E.: Named entities: recognition, classification and use. John Benjamins Publishing (2009)
22. Garrido, A.L., Sangiao, S., Cardiel, O.: Improving the generation of infoboxes from data silos through machine learning and the use of semantic repositories. Int. J. Artif. Intell. Tools **26**(05), 1760022 (2017)

23. Garrido, A.L., Gomez, O., Ilarri, S., Mena, E.: NASS: news annotation semantic system. In: Proceedings of the IEEE International Conference on Tools with Artificial Intelligence, pp. 904–905. IEEE (2011)
24. Buey, M.G., Garrido, A.L., Escudero, S., Trillo, R., Ilarri, S., Mena, E.: SQX-Lib: developing a semantic query expansion system in a media group. In European Conference on Information Retrieval, pp. 780–783. Springer, Heidelberg (2014)
25. Garrido, A. L., Gómez, O., Ilarri, S., Mena, E.: An experience developing a semantic annotation system in a media group. In: Proceedings of the International Conference on Application of Natural Language to Information Systems, pp. 333–338. Springer, Heidelberg (2011)
26. Garrido, A.L., Ilarri, S., Mena, E.: GEO-NASS: a semantic tagging experience from geographical data on the media. In: Proceedings of the East European Conference on Advances in Databases and Information Systems, pp. 56–69. Springer, Heidelberg (2013)
27. Lin, J.: The lambda and the kappa. IEEE Internet Comput. **21**(5), 60–66 (2017)

A Conversion of Feature Models into an Executable Representation in Microsoft Excel

Viet-Man Le(✉), Thi Ngoc Trang Tran, and Alexander Felfernig

Graz University of Technology, Graz, Austria
{vietman.le,ttrang,alexander.felfernig}@ist.tugraz.at

Abstract. Feature model-based configuration involves selecting desired features from a collection of features (called a feature model) that satisfy pre-defined constraints. Configurator development can be performed by different stakeholders with distinct skills and interests, who could also be non-IT domain experts with limited technical understanding and programming experience. In this context, a simple configuration framework is required to facilitate non-IT stakeholders' participation in configurator development processes. In this paper, we develop a so-called tool FM2ExConf that enables stakeholders to represent configuration knowledge as an executable representation in Microsoft Excel. Our tool supports the conversion of a feature model into an Excel-based configurator, which is performed in two steps. In the first step, the tool checks the consistency and anomalies of a feature model. If the feature model is consistent, then it is converted into a corresponding Excel-based configurator. Otherwise, the tool provides corrective explanations that help stakeholders to resolve anomalies before performing the conversion. Besides, in the second step, another type of explanation (which is included in the Excel-based configurator) is provided to help non-IT stakeholders to fix inconsistencies in the configuration phase.

Keywords: Feature models · Knowledge-based configuration · Knowledge acquisition · Configurator · Microsoft Excel · Automated analyses · Anomalies · Explanations

1 Introduction

Knowledge-based configuration encompasses all activities related to the configuration of products from predefined components while respecting a set of well-defined constraints that restrict infeasible products [18]. Configuration has been applied in various domains, such as *financial services* [10], *requirements engineering* [19], *telecommunication* [12], and *furniture industry* [14]. In configuration systems, knowledge bases often play a crucial role in reflecting the real-world product domain. Many communication iterations between domain

M. Stettinger et al. (Eds.): ISMIS 2020, SCI 949, pp. 153–168, 2021.
https://doi.org/10.1007/978-3-030-67148-8_12

experts and knowledge engineers are necessary to develop and maintain a configuration knowledge base. In this context, *feature models* [16] have been recognized as conventional means to facilitate the collaborative model development. Like UML-based configuration models [8], feature models provide a graphical representation that improves the understandability of knowledge bases and the efficiency of underlying development processes [7]. Moreover, these models help stakeholders to decide on relevant features and learn about existing dependencies between features.

Microsoft Excel[1] has been recognized as one of the most widely used spreadsheet applications in modern society. This tool enables non-programmers to perform programming-like tasks in a visual tabular approach. In the current literature, there exist several studies (e.g., [3,9]) that leverage Excel to tackle configuration problems. The popularity and usability of Excel motivate us to come up with the idea of using this tool to support the configurator development process of non-IT stakeholders.

On the other hand, real-world feature models usually consist of a vast number of features and variants. The magnitude and the inherent complexity of constraints in the feature models can trigger latent *anomalies*, which become manifest in different types of *inconsistencies* and *redundancies* [6]. In this context, developing tools that help to identify such feature model anomalies has become crucial to avoid burdens concerning feature model development and maintenance. In this paper, we develop a so-called tool FM2ExConf with two key functionalities: (1) detecting and explaining feature model anomalies, and (2) converting a consistent feature model into an Excel-based configurator. The detection of anomalies is performed based on an approach presented in [6], meanwhile, the anomaly explanations are generated on the basis of two algorithms - FastDiag and FMCore [6,9]. These algorithms have been proven effective in generating minimal corrective explanations as *diagnoses*[2] for feature model anomalies. Regarding the conversion of feature model into an Excel-based configurator, we propose an approach that utilizes Excel worksheets as a complementary means to model configuration knowledge on the basis of feature model concepts. Besides, we introduce a method using Excel formula to generate corrective explanations, which are helpful for non-IT stakeholders to resolve inconsistencies in the configuration phase.

The remainder of the paper is structured as follows. A brief revisit of feature model-based configuration is presented in Sect. 2. In Sect. 3, we present the architecture of the tool and show how it helps to detect the anomalies of a feature model (Subsect. 3.2) and how it supports the conversion of the feature model into an Excel-based configurator (Subsect. 3.3). Related work is summarized in Sect. 4 and a discussion on the pros and cons of the presented approach as well as an outlook in terms of future work are presented in Sect. 5. Finally, the paper is concluded in Sect. 6.

1 www.office.com.

2 *A diagnosis* is a minimal set of constraints, which have to be adapted or deleted from an inconsistent feature model such that the remaining constraints allow the calculation of at least one configuration [6].

2 Feature Model-Based Configuration

2.1 Definitions

In feature modeling, feature models represent all possible configurations of a configuration task in terms of *features* and their *interrelationships* [2,16]. Features are organized hierarchically as a tree structure, where *nodes* represent the features, and *links* represent relationships between nodes. *Features* and *relationships* are equivalent to the *variables* and *constraints* of a CSP[3]-based configuration task [15]. Each variable f_i has a specified domain $d_i = \{true, false\}$. An example feature model of a "*Bamboo Bike*", which is inspired by products of *my Boo* brand[4], is depicted in Fig. 1. The detailed description of this model is presented in Subsect. 2.2.

For the following discussions, we introduce the definitions of a *feature model configuration task* and a *feature model configuration* (solution) [6,15].

Definition 1 (Feature model configuration task). *A feature configuration task is defined by a triple* $(\boldsymbol{F}, \boldsymbol{D}, \boldsymbol{C})$*, where* $\boldsymbol{F} = \{f_1, f_2, ..., f_n\}$ *is a set of features,* $\boldsymbol{D} = \{dom(f_1), dom(f_2), ..., dom(f_n)\}$ *is the set of feature domains, and* $\boldsymbol{C} = CF \cup CR$ *is a set of constraints restricting possible configurations,* $CF = \{c_1, c_2, ..., c_k\}$ *represents a set of feature model constraints, and* $CR = \{c_{k+1}, c_{k+2}, ..., c_m\}$ *represents a set of user requirements.*

Definition 2 (Feature model configuration). *A feature model configuration* $\boldsymbol{S}$ *for a given feature model configuration task* (F, D, C) *is an* assignment *of the features* $f_i \in F, \forall i \in [1..n]$*.* S *is* **valid** *if it is* complete *(i.e., each feature in* F *has a value) and* consistent *(i.e.,* S *fulfills the constraints in* C*).*

Based on the aforementioned definitions, in the next subsection, we introduce feature model concepts[5], which are commonly applied to specify configuration knowledge [15]. Besides, we exemplify a Bamboo Bike feature model (see Fig. 1) to explain the concepts.

2.2 Feature Model Concepts

A feature model (configuration model) consists of two parts: *structural part* and *constraint part*. The former establishes a hierarchical relationship between features. The latter combines additional constraints that represent so-called *cross-tree constraints*.

Structurally, a feature model is a rooted tree, where nodes are features. Each feature is identified by a unique *name*, which exploited to describe possible *states* of a feature (i.e., "*included in*" or "*excluded from*" a specific configuration) [15].

[3] CSP - Constraint Satisfaction Problem.
[4] www.my-boo.com.
[5] For further model concepts, we refer to [1,2].

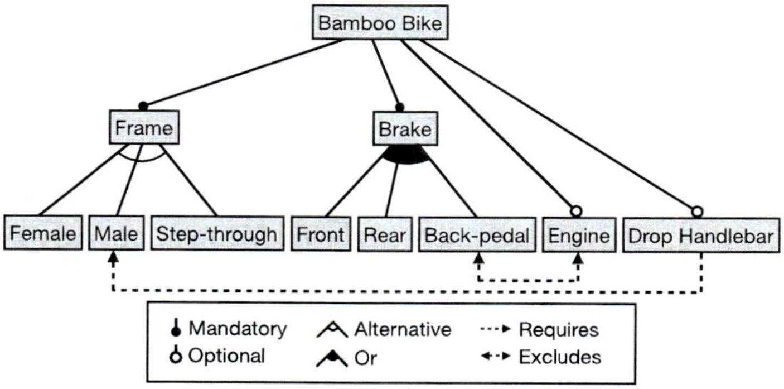

Fig. 1. A simplified feature model of the Bamboo Bike Configuration. The *"Step-through"* feature describes the brand-new bamboo frame from *my Boo* brand.

The root of the tree is a so-called *root feature* f_r, which is involved in every configuration ($f_r = true$). Besides, each feature can have other features as its *subfeatures*. The relationship between a feature and its subfeatures can be typically classified as follows:

- *Mandatory relationship*: A *mandatory* relationship between two features f_1 and f_2 indicates that f_2 will be included in a configuration *if and only if* f_1 is included in the configuration. For instance, in Fig. 1, `Frame` and `Brake` show mandatory relationships with `Bamboo Bike`. Since `Bamboo Bike` is the root feature, `Frame` and `Brake` must be included in all configurations.
- *Optional relationship*: An *optional* relationship between two features f_1 and f_2 indicates that if f_1 is included in a configuration, then f_2 *may* or *may not* be included in the configuration. In Fig. 1, the relationship between `Bamboo Bike` and `Engine` is *optional*.
- *Alternative relationship*: An *alternative* relationship between a feature f_p and its subfeatures $C = \{f_1, f_2, ..., f_k\} (C \in F)$ indicates that if f_p is included in a configuration, then *exactly one* $f_c \in C$ must be included in the configuration. For instance, in Fig. 1, the relationship between `Frame` and its subfeatures (`Female`, `Male`, and `Step-through`) is *alternative*.
- *Or relationship*: An *or* relationship between a feature f_p and its subfeatures $C = \{f_1, f_2, ..., f_k\} (C \in F)$ indicates that if f_p is included in a configuration, then *at least one* $f_c \in C$ must be included in the configuration. For instance, in Fig. 1, the relationship between `Brake` and its subfeatures (`Front`, `Rear`, and `Back-pedal`) reflects an *or* relationship.

In the *constraint part*, additional constraints are integrated graphically into the model to set cross-hierarchical restrictions for features. According to [15], the following constraint types are used for the specification of feature models:

- *Requires*: A *requires* constraint between two features (*"f_1 requires f_2"*) indicates that if feature f_1 is included in the configuration, then f_2 must also be

included. For instance, in Fig. 1, if a `Drop Handlebar` is included in a configuration, then a `Male` frame must be included as well. The dashed line directed from `Drop Handlebar` to `Male` denotes a *requires* constraint.
- *Excludes*: An *excludes* constraint between two features ("f_1 *excludes* f_2") indicates that both f_1 and f_2 must not be included in the same configuration. For instance, in Fig. 1, `Engine` must not be combined with a `Back-pedal` brake. The dashed line with two arrows between `Engine` and `Back-pedal` denotes an *excludes* constraint.

The mentioned relationships and constraints can be translated into a static CSP representation using the rules in Table 1.

Table 1. Semantics of feature model concepts in static CSPs (P, C, C_i, A, and B represent individual features).

Relationship/Constraint	Semantic in static CSP
$mandatory(P, C)$	$P \leftrightarrow C$
$optional(P, C)$	$C \rightarrow P$
$or(P, C_1, C_2, \ldots, C_n)$	$P \leftrightarrow (C_1 \vee C_2 \vee \ldots \vee C_n)$
$alternative(P, C_1, C_2, \ldots, C_n)$	$(C_1 \leftrightarrow (\neg C_2 \wedge \ldots \wedge \neg C_n \wedge P))$ $\wedge(C_2 \leftrightarrow (\neg C_1 \wedge \neg C_3 \wedge \ldots \wedge \neg C_n \wedge P))$ $\wedge \ldots$ $\wedge(C_n \leftrightarrow (\neg C_1 \wedge \ldots \wedge \neg C_{n-1} \wedge P))$
$requires(A, B)$	$A \rightarrow B$
$excludes(A, B)$	$\neg A \vee \neg B$

3 Convert a Feature Model into an Excel-Based Configurator

To facilitate the participation of non-IT stakeholders in configurator development processes, we develop a tool (called FM2EXCONF) that supports the conversion of a feature model into an Excel-based configurator. In the following subsections, we present the tool's architecture as well as its key components.

3.1 Fm2ExConf Architecture

FM2EXCONF consists of two key components: (1) *anomaly detection and explanation*, which identifies anomalies (in terms of *inconsistencies* and *redundancies*) of a feature model and generates corrective explanations to resolve the anomalies, and (2) *feature model conversion*, which converts a consistent feature model into an Excel-based configurator. The input of the tool is a *feature model file*, and the output is an *Excel file* representing a corresponding Excel-based configurator. The tool supports the following formats of feature model files:

- SXFM format - which is used to encode feature models in S.P.L.O.T.'s web application [17]. We use the available S.P.L.O.T.'s Java parser library to read SXFM files.
- FEATUREIDE XML format - presented by FEATUREIDE [23] and used in FEATUREIDE plugin. We use Java DOM Parser to read the file.
- GLENCOE JSON format - which is used to encode feature models in GLENCOE web application [22]. We use JSON decoder of ORG.JSON library to convert the JSON format into Java objects.
- *Descriptive* format - which uses a simple and descriptive representation of relationships/constraints presented in [15]. An example of the representation of this format can be found in the left part of Fig. 3, the item *"Details"*.

The tool transforms the feature model file into a Java object that could be understood by the FM2EXCONF's engine. Figure 2 shows the architecture and the key functionalities of the tool, which are described in the following subsections.

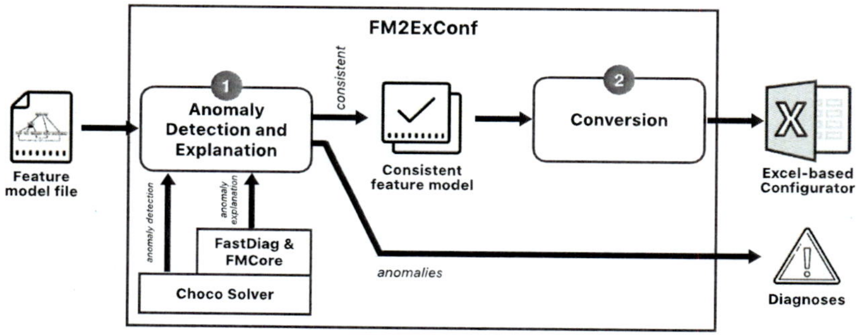

Fig. 2. The architecture (In Fig. 2, we use the icons from https://icons8.com, including *General Warning Sign* icon, File icon, *Microsoft Excel* icon, *Check All* icon.) of the FM2EXCONF showing two key components. The first component detects anomalies of a feature model and generates minimal explanations (diagnoses) for resolving anomalies. The second component is responsible for converting the feature model into an Excel-based configurator.

3.2 Detecting and Explaining Feature Model Anomalies

Due to the increasing size and complexity of feature models, anomalies in terms of inconsistencies and redundancies can occur [6]. To avoid the generation of configurators from feature models involving anomalous features or redundant constraints, our tool enables an automated analysis process to identify feature model anomalies. The process is performed based on the approach proposed by Felfernig et al. [6]. A conversion of a feature model into an Excel-based configuration (see Subsect. 3.3) can only be done if the feature model is consistent. The analysis process is executed in two steps: *anomaly detection* and *anomaly explanation generation*. Figure 3 shows an example of how a Bamboo Bike feature model is analyzed to identify anomalies and corresponding explanations. This is a modified version of the Bamboo Bike feature model presented in Fig. 1 with an additional constraint *requires(Brake, Male)*.

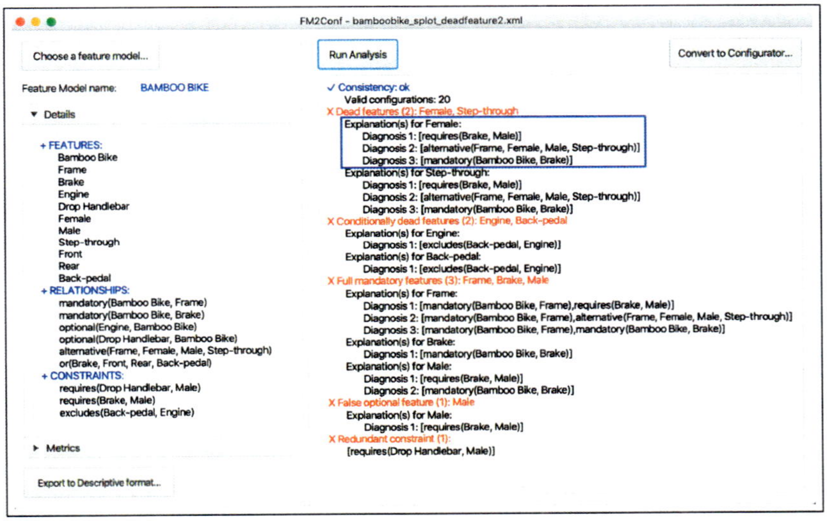

Fig. 3. The user interface of FM2ExConf showing how a feature model in the Bamboo Bike domain can be analyzed to determine anomalies and corrective explanations. The user interface includes two parts. The left part allows a user to load a feature model file. The user is also able to review the feature model in the descriptive format and to see a feature model statistics in the item *"Metrics"*, such as *the number of features, relationships*, or *constraints*. The right part shows two buttons corresponding to two key functionalities: (1) *"Run Analysis"*, which proceeds a feature model analysis and shows the results in the text area, and (2) *"Convert to Configurator"*, which generates a configurator in an Excel worksheet. This functionality is only activated when the feature model is consistent (i.e., the analysis result shows *"Consistency: ok"*).

- *Anomaly detection*: The tool applies the checking methods proposed by Felfernig et al. [6] to detect six types of feature model anomalies: *void feature models, dead features, conditionally dead features, full mandatory features, false optional features,* and *redundant constraints. A void feature model* is a feature model that represents no configurations. *A dead feature* is a feature that is not included in any of the possible configurations. *A conditionally dead feature* is a feature that becomes dead under certain circumstances (e.g. when including another feature(s) in a configuration). *A full mandatory feature* is

a feature that is included in every possible solution. *A false optional feature* is a feature that is included in all configurations, although it has not been modeled as mandatory. *A redundant constraint* is a constraint whose semantic information has already been modeled in another way in other constraints/relationships of the feature model. For further details of anomalies, we refer to [2,6].
- *Anomaly explanation generation*: The tool uses FastDiag and FMCore algorithms to generate corrective explanations, which help stakeholders to resolve anomalies of the feature model. Since FastDiag determines exactly one diagnosis at a time, we combined this algorithms FastDiag with a construction of the hitting set directed acyclic graph (HSDAG) introduced by Reiter [20] in order to determine the complete set of diagnoses.[6]

An example explanation can be found in Fig. 3 (see the explanation covered by the blue rectangle). The tool detects that the feature `Female` is a *dead feature* and generates three corrective explanations: *Diagnosis 1: requires(Brake, Male)*, *Diagnosis 2: alternative(Frame, Female, Male, Step-through)*, and *Diagnosis 3: mandatory(Bamboo Bike, Brake)*. These explanations represent three ways to delete/adapt the relationships/constraints in the feature model to resolve the mentioned anomaly. Particularly, the dead feature `Female` can be resolved if the stakeholder deletes/adapts either the constraint *"requires"* between `Brake` and `Male`, or the *"alternative"* relationship between `Frame` and its sub-features (`Female, Male, Step-through`), or the *"mandatory"* relationship between `Bamboo Bike` and `Brake`.

3.3 Convert a Feature Model into an Excel-Based Configurator

After the consistency check, the conversion step is executed to convert the feature model into an Excel-based configurator. In this subsection, we introduce an approach to represent a feature model in an Excel worksheet. Besides, we also present how to use our tool to generate an Excel-based configurator.

a. Represent a Feature Model in an Excel Worksheet. Our approach is to utilize an Excel worksheet to represent feature models, for both structural and constraint parts. An Excel worksheet represents three elements of a feature model: (1) *names*, (2) *states*, and (3) *relationships/constraints*. The *names* represent the structure of a feature model. The *states* store the current state of features in a specific configuration (e.g., "*included*"/"*excluded*"). The *relationships/constraints* are represented in two forms. First, *text-based rules* are exploited to enable stakeholders to understand the relationship between features. Second, *Excel formulae* are used to generate corrective explanations that help stakeholders to resolve configuration inconsistencies. These formulae are translated from the relationships/constraints using logical test functions.

[6] For further details of combining FastDiag with a construction of HSDAG, we refer to [6,11].

	A	B	C	D	E
1	FEATURE	INCLUDED (1 yes, 0 no)	RELATIONSHIP/CONSTRAINT	OK	PRICE
2	Bamboo Bike	1			950
3	Frame	1	Bamboo Bike <-> Frame	ok	0
4	Brake	0	Bamboo Bike <-> Brake	*include Brake*	0
5	Engine	1	Engine -> Bamboo Bike	ok	200
6	Drop Handlebar	0	Drop Handlebar -> Bamboo Bike	ok	0
7	Frame		Frame -> XOR(Female, Male, Step-through)	*include 1 out of Female, Male, Step-through*	
8	Female	1	Female -> Frame	ok	300
9	Male	0	Male -> Frame	ok	0
10	Step-through	1	Step-through -> Frame	ok	400
11	Brake		Brake -> OR(Front, Rear, Back-pedal)	*include Brake or exclude Brake's subfeatures*	
12	Front	1	Front -> Brake	*exclude Front or include Brake*	50
13	Rear	0	Rear -> Brake	ok	0
14	Back-pedal	0	Back-pedal -> Brake	ok	0
15	Cross-Tree Constraints		Drop Handlebar -> Male	ok	
16			not(Back-pedal) or not(Engine)	ok	

Fig. 4. An Excel-based configurator for the Bamboo Bike feature model (see Fig. 1) generated by FM2EXCONF. The features of this model are listed in breadth-first order.

The conversion of a feature model into an executable representation in an Excel worksheet can be conducted in the following steps:

- **Step 1**: Put feature names in the **first column**. The features conform to one of the following orders:
  - *Breadth-first order*: The list of feature names is retrieved by traversing level-by-level in the feature model. The process starts with the root feature f_r, then comes to the subfeatures of the root feature before moving to other features at the next level. This process is repeated until the final level is reached.
  - *Depth-first order*: The list of feature names is retrieved by traversing the feature model in a depth-first fashion. The list starts with the root feature f_r, then follows the path of corresponding subfeatures as far as it can go (i.e., from the root feature to its leaf features). The process continues until the entire graph has been traversed.

- **Step 2**: Reserve cells in the **second column** to save the *states* of the features. The cells will be filled in the configuration phase when users manually change the cells' value to find configurations. The value of cells is binary values (`1/0`) or logical values (`TRUE/FALSE`), which represent two states of a feature ("*included*"/"*excluded*").
- **Step 3**: Fill the **third column** with *text-based rules* that represent the relationships/constraints between features. The text-based rules can be represented according to relationship/constraint types as follows:
  - *Mandatory* and *Optional*: Each relationship is placed in the row of the feature that participates in the relationship (see cells $C3$–$C6$ in Fig. 4). The feature is in the left part of the rule (except for a *mandatory* relationship, where the feature is in the right part).

- *Alternative* and *Or*: Insert a new row above the subfeatures to store the relationship (see rows 7&11 in Fig. 4). For instance, to represent an *alternative* relationship between `Frame` and its subfeatures (`Female`, `Male`, `Step-through`), we insert a new row above the subfeatures (see row 7 in Fig. 4), and in cell $C7$, we add a corresponding rule of the *alternative* relationship. Besides, for each subfeature, add a *requires* constraint to check the consistency between the subfeature and its parentfeature (see cells $C8$–$C10$, $C12$–$C14$ in Fig. 4).
- *Constraints* are located at the end of the relationship list (see constraints in rows 15 & 16 in Fig. 4).

– **Step 4**: Convert relationships/constraints into logical test formulae and save in the **fourth column**. The returns of these formulae are used as *textual explanations* that describe the consistency of feature assignments or suggest corrective solutions when inconsistencies occur. Tables 2, 3, 4, 5 and 6 provide formula templates to generate such explanations according to *six* relationship/constraint types. The templates are derived from truth tables [21], where the last column shows how an explanation can be formulated (e.g., *"ok"* if consistent, *"include feature A"* if inconsistent).

Besides textual explanations, *visual explanations* are exploited to graphically represent warnings concerning the inconsistency of the corresponding relationships/constraints. In Excel, the warnings can be created using conditional formatting. For instance, in our example, we set conditional formatting to color a cell in column D with light red if the formula of this cell does not return the string *"ok"* (see cells $D4$, $D7$, $D11$ & $D12$ in Fig. 4).

– **Step 5**: Integrate services such as *pricing* and *capacity* of the product configuration domain into the **remaining columns**. For instance, in our example, the *fifth column* shows the price of each feature and the total price of a configuration (see Fig. 4).

Table 2. The truth table and the derived Excel formula template for *Mandatory* relationships.

A	B	$A \leftrightarrow B$	Explanation
0	0	1	ok
0	1	0	include A
1	0	0	include B
1	1	1	ok

Derived Excel formula template:

```
=IF(A_ref=0,IF(B_ref=1,''*include A*'',''ok''),
IF(B_ref=0,''*include B*'',''ok''))
```

Table 3. The truth table and the derived Excel formula template for *Optional* relationships and *Requires* constraints.

A	B	$A \rightarrow B$	Explanation
0	0	1	ok
0	1	1	ok
1	0	0	exclude A or include B
1	1	1	ok

Derived Excel formula template:

```
=IF(A_ref=1,IF(B_ref=0,''*exclude A or include B*'',''ok''),''ok'')
```

Table 4. The truth table and the derived Excel formula template for *Or* relationships.

A	B	C	$A \leftrightarrow (B \vee C)$	Explanation
0	0	0	1	ok
0	0	1	0	include A or exclude A's subfeatures
0	1	0	0	include A or exclude A's subfeatures
0	1	1	0	include A or exclude A's subfeatures
1	0	0	0	include B or C
1	0	1	1	ok
1	1	0	1	ok
1	1	1	1	ok

Derived Excel formula template:

```
=IF(B_ref+C_ref=0,IF(A_ref=1,''*include B or C*'',''ok''),
IF(A_ref=0,''*include A or exclude A's subfeatures*'',''ok''))
```

Table 5. The truth table and the derived Excel formula template for *Alternative* relationships.

A	B	C	$(B \leftrightarrow (\neg C \wedge A)) \wedge (C \leftrightarrow (\neg B \wedge A))$	Explanation
0	0	0	1	ok
0	0	1	0	include A
0	1	0	0	include A
0	1	1	0	include 1 out of B, C
1	0	0	0	include 1 out of B, C
1	0	1	1	ok
1	1	0	1	ok
1	1	1	0	include 1 out of B, C

Derived Excel formula template:

```
=IF(B_ref+C_ref=1,IF(A_ref=0,''*include A*'',''ok''),
IF(A_ref+B_ref+C_ref=0,''ok'',''*include 1 out of B, C*''))
```

Table 6. The truth table and the derived Excel formula template for *Excludes* constraints.

A	B	$\neg A \vee \neg B$	Explanation
0	0	1	ok
0	1	1	ok
1	0	1	ok
1	1	0	exclude A or B
Derived Excel formula template:			
`=IF(A_ref=1,IF(B_ref=1,''*exclude A or B*'',''ok''),''ok'')`			

b. Generate an Excel-Based Configurator. After parsing an input file into a feature model object, FM2ExConf performs all the steps mentioned in Subsect. 3.3*a* to generate an Excel worksheet (.xlsx), which stores a corresponding configurator. Besides, the tool converts the relationships/constraints of the feature model into three forms: *text-based rules, Excel formulae,* and *Choco's constraints.* Text-based rules and Excel formulae have been mentioned in Subsect. 3.3*a*. Choco's constraints are exploited to detect and explain feature model anomalies (see Subsect. 3.2).

To save the generated configurator in an Excel file, we use the library *Apache POI* - a well-known Java API for Microsoft Documents, which is able to read and write Excel files. Apache POI provides functions to adapt Excel file formats, such as customizing font styles, changing the background, building formulae, and setting conditional formatting.

Besides, as mentioned in Subsect. 3.3*a*, FM2ExConf allows users to set options for a configurator, e.g., *the order of features in the first column* (breadth-first order or depth-first order) or *the state of a feature* (binary values (`1/0`) or logical values (`TRUE/FALSE`)). Figure 4 shows an Excel-based configurator converted from a Bamboo Bike feature model (see Fig. 1). In this configurator, the features are listed in breadth-first order, and binary values (`1/0`) are used to represent the state of features (*"included"/"excluded"*).

4 Related Work

Studies related to our work are categorized into two groups. The first group involves studies that leverage Excel to tackle configuration problems. The second group includes feature model tools that enable automatic analysis operations on feature models.

The utilization of Excel to tackle configuration problems are proposed by Felfernig et al. [9], and Bordeaux et al. [3]. These approaches are similar to ours with regard to the application scenario, where a user can interactively select the desired features and immediately get feedback on incompatible choices. However, these approaches focus on the programmatic integration of an underlying constraint solver into Excel and require the knowledge regarding constraint satisfaction concepts to design and to use the system, which could be a bit challenging to non-IT users. Our approach tries to steer clear of this issue by generating Excel-based configurators based on feature model concepts, which are easy to understand and widely applied to manage product variants. This way, our approach is feasible for different stakeholders, ranging from knowledge engineers to end-users.

Regarding automatic analysis operations on feature models, the current literature has shown plenty of tools that support feature model creation and analysis, such as FEATUREIDE [23], S.P.L.O.T. [17], and GLENCOE [22].

These tools enable to create feature models, in which the feature addition is done by direct interactions on the feature model tree, i.e., right-click on a feature to create its subfeature. Besides, to represent cross-tree constraints of feature models, GLENCOE [22] allows to draw curved and directed lines, which increase the understandability of the feature model. Thus, to take advantage of such good support for feature model creation, FM2EXCONF takes feature model files created by these tools as the input but does not provide a feature model creation functionality.

Besides, these tools provide various types of support for anomaly detection and explanation. For instance, S.P.L.O.T. checks consistency and detects dead or full mandatory features. FEATUREIDE and GLENCOE support the detection of all anomaly types. To resolve anomalies, while S.P.L.O.T. only marks and lists dead and full mandatory features, GLENCOE and FEATUREIDE can highlight anomalous features and constraints that could trigger inconsistencies in the feature model. Especially, FEATUREIDE can additionally generate explanations in a user-friendly manner.

Like FEATUREIDE and GLENCOE, FM2EXCONF can detect all types of feature model anomalies and provide corrective explanations if anomalies exist in the feature model. Differ from other tools, FM2EXCONF creates explanations that are represented in minimal sets of constraints (diagnoses). This way, stakeholders can easily detect relationships/constraints that contribute mainly to anomalies.

5 Discussion

Our first experiments have shown that Excel-based configurators, which are generated by our tool, can facilitate the participation of non-IT stakeholders in configuration development processes. In particular, the usage of Excel's spreadsheet interface paradigm helps to maintain the most important benefits of feature models (i.e., the *feature hierarchy*) and provides stakeholders with an overview of product variants. Besides, the formulation of corrective explanations in Excel-based configurators enables non-IT stakeholders to resolve configuration inconsistencies in the configuration phase. Thereby, Excel-based configurators can be exploited to reduce efforts and risks related to configuration knowledge acquisition.

Besides, our approach, which represents configuration knowledge as an executable representation in Excel, is also applicable in other spreadsheet programs such as *Numbers* and *OpenOffice Calc*. The representation is quite straightforward and appropriate for in-complex feature models, and therefore helpful for small and medium-sized enterprises to overcome challenges concerning configurator implementation and utilization (e.g., high costs or considerable chances of failure [13]).

Moreover, a mechanism that allows to manually select/deselect features in an Excel worksheet to find configurations might provide users with a simulation of how a configuration task is done. Thus, Excel-based configurators can be applied in several scenarios, such as (1) facilitating the participation of non-IT stakeholders in knowledge engineering processes, (2) including customers in open innovation processes, and (3) enabling trainers to give learners easy hands-on experiences with feature modeling.

Our proposal has three limitations that need to be improved within the scope of future work. First, the anomaly explanations of our tool are given as constraint sets, which are not directly related to the feature model's structural information. This could be a challenge for stakeholders to comprehend anomalies. Thus, a means to express explanations in a user-friendly manner is required. The second limitation lies in the built-in reasoning engine of Excel (i.e., Excel solver). The Excel solver requires certain knowledge to set up necessary parameters to find solutions, which could be a bit challenging to end-users. Besides, the Excel solver is able to find only one configuration at once instead of a set of configurations. Hence, the utilization of constraint-based solving add-ons presented in [4,9] can be a potential solution to resolve this issue of our approach. Finally, the conversion of a feature model into an Excel-based configurator currently only supports basic feature model concepts [16]. An extension to support cardinality-based feature models [5] and extended feature models [1] are therefore necessary to make our approach more applicable for real-world feature models.

6 Conclusion

In this paper, we developed a so-called tool FM2EXCONF that allows stakeholders to represent configuration knowledge as an executable representation in Microsoft Excel. Our tool enables two main functionalities. The first functionality supports anomaly detection and anomaly explanation generation, which help stakeholders to identify and resolve anomalies in a feature model. The generation of corrective explanations is performed based on two algorithms - FASTDIAG and FMCORE. The second functionality is to convert a consistent feature model into an Excel-based configurator. To support the second functionality, we proposed a novel approach that provides a guideline on the representation of a basic feature model in an Excel worksheet. Besides, we presented throughout the paper example feature models in the Bamboo Bike domain to illustrate our approach. Although our approach facilitates the configurator development process of stakeholders, some improvements concerning corrective explanations should be proceeded within the scope of future work in order to provide more user-friendly explanations to stakeholders.

References

1. Batory, D.: Feature models, grammars, and propositional formulas. In: Obbink, H., Pohl, K. (eds.) International Conference on Software Product Lines, pp. 7–20. Springer, Heidelberg (2005). https://doi.org/10.1007/11554844_3
2. Benavides, D., Segura, S., Ruiz-Cortés, A.: Automated analysis of feature models 20 years later: a literature review. Inf. Syst. **35**(6), 615–636 (2010). https://doi.org/10.1016/j.is.2010.01.001
3. Bordeaux, L., Hamadi, Y.: Solving configuration problems in excel. In: Proceedings of the 2007 AAAI Workshop, pp. 38–40. Configuration, The AAAI Press, Menlo Park, California (2007)
4. Chitnis, S., Yennamani, M., Gupta, G.: ExSched: solving constraint satisfaction problems with the spreadsheet paradigm. In: 16th Workshop on Logic-Based Methods in Programming Environments (WLPE2006) (2007)
5. Czarnecki, K., Helsen, S., Eisenecker, U.: Formalizing cardinality-based feature models and their specialization. Softw. Process Improv. Pract. **10**(1), 7–29 (2005). https://doi.org/10.1002/spip.213
6. Felfernig, A., Benavides, D., Galindo, J., Reinfrank, F.: Towards anomaly explanation in feature models. In: ConfWS-2013: 15th International Configuration Workshop (2013), vol. 1128, pp. 117–124 (Aug 2013)
7. Felfernig, A.: Standardized configuration knowledge representations as technological foundation for mass customization. IEEE Trans. Eng. Manag. **54**(1), 41–56 (2007). https://doi.org/10.1109/TEM.2006.889066
8. Felfernig, A., Friedrich, G., Jannach, D.: UML as domain specific language for the construction of knowledge-based configuration systems. Int. J. Softw. Eng. Knowl. Eng. **10**(04), 449–469 (2000). https://doi.org/10.1142/s0218194000000249
9. Felfernig, A., Friedrich, G., Jannach, D., Russ, C., Zanker, M.: Developing constraint-based applications with spreadsheets. In: Chung, P.W.H., Hinde, C., Ali, M. (eds.) Developments in Applied Artificial Intelligence. IEA/AIE 2003, vol. 2718, pp. 197–207. Springer, Heidelberg (2003). https://doi.org/10.1007/3-540-45034-3_20
10. Felfernig, A., Isak, K., Szabo, K., Zachar, P.: The VITA financial services sales support environment. In: Proceedings of the 19th National Conference on Innovative Applications of Artificial Intelligence - Volume 2, pp. 1692–1699. IAAI'07, AAAI Press (2007). https://doi.org/10.5555/1620113.1620117
11. Felfernig, A., Schubert, M., Zehentner, C.: An efficient diagnosis algorithm for inconsistent constraint sets. Artif. Intell. Eng. Des. Anal. Manuf. **26**(1), 53–62 (2012). https://doi.org/10.1017/S0890060411000011
12. Fleishanderl, G., Friedrich, G.E., Haselbock, A., Schreiner, H., Stumptner, M.: Configuring large systems using generative constraint satisfaction. IEEE Intell. Syst. Appl. **13**(4), 59–68 (1998). https://doi.org/10.1109/5254.708434
13. Forza, C., Salvador, F.: Product Information Management for Mass Customization: Connecting Customer, Front-Office and Back-Office for Fast and Efficient Customization. Palgrave Macmillan, London (2006). https://doi.org/10.1057/9780230800922

14. Haag, A.: Sales configuration in business processes. IEEE Intell. Syst. Appl. **13**(4), 78–85 (1998). https://doi.org/10.1109/5254.708436
15. Hotz, L., Felfernig, A., Stumptner, M., Ryabokon, A., Bagley, C., Wolter, K.: Chapter 6 - configuration knowledge representation and reasoning. In: Felfernig, A., Hotz, L., Bagley, C., Tiihonen, J. (eds.) Knowledge-Based Configuration, pp. 41–72. Morgan Kaufmann, Boston (2014). https://doi.org/10.1016/B978-0-12-415817-7.00006-2
16. Kang, K., Cohen, S., Hess, J., Novak, W., Peterson, A.: Feature-Oriented Domain Analysis (FODA) Feasibility Study. Technical Report. CMU/SEI-90-TR-021, Software Engineering Institute, Carnegie Mellon University, Pittsburgh, PA (1990). http://resources.sei.cmu.edu/library/asset-view.cfm?AssetID=11231
17. Mendonca, M., Branco, M., Cowan, D.: S.P.L.O.T.: software product lines online tools. In: Proceedings of the 24th ACM SIGPLAN Conference Companion on Object Oriented Programming Systems Languages and Applications, pp. 761–762. OOPSLA'09. ACM, New York (2009). https://doi.org/10.1145/1639950.1640002
18. Mittal, S., Frayman, F.: Towards a generic model of configuration tasks. In: Proceedings of the 11th International Joint Conference on Artificial Intelligence - Volume 2, pp. 1395–1401. IJCA'89, Morgan Kaufmann Publishers Inc., San Francisco, CA, USA (1989). https://doi.org/10.5555/1623891.1623978
19. Ninaus, G., Felfernig, A., Stettinger, M., Reiterer, S., Leitner, G., Weninger, L., Schanil, W.: INTELLIREQ: intelligent techniques for software requirements engineering. In: Proceedings of the Twenty-First European Conference on Artificial Intelligence, pp. 1161–1166. ECAI'14, IOS Press, NLD (2014). https://doi.org/10.5555/3006652.3006911
20. Reiter, R.: A theory of diagnosis from first principles. Artif. Intell. **32**(1), 57–95 (1987). https://doi.org/10.1016/0004-3702(87)90062-2
21. Rosen, K.H.: Discrete Mathematics and Its Applications, 5th edn. McGraw-Hill Higher Education, New York (2002)
22. Schmitt, G.R.A., Bettinger, C., Rock, G.: Glencoe–a tool for specification, visualization and formal analysis of product lines. In: Proceedings of ISTE 25th International Conference on Transdisciplinary Engineering. Advances in Transdisciplinary Engineering, vol. 7, pp. 665–673. IOS Press, Amsterdam (2018). https://doi.org/10.3233/978-1-61499-898-3-665
23. Thüm, T., Kästner, C., Benduhn, F., Meinicke, J., Saake, G., Leich, T.: FeatureIDE: an extensible framework for feature-oriented software development. Sci. Comput. Program. **79**, 70–85 (2014). https://doi.org/10.1016/j.scico.2012.06.002

Basic Research and Algorithmic Problems

Explainable Artificial Intelligence. Model Discovery with Constraint Programming

Antoni Ligęza(✉), Paweł Jemioło, Weronika T. Adrian, Mateusz Ślażyński, Marek Adrian, Krystian Jobczyk, Krzysztof Kluza, Bernadetta Stachura-Terlecka, and Piotr Wiśniewski

AGH University of Science and Technology,
al. A.Mickiewicza 30, 30-059 Krakow, Poland
{ligeza,pawel.jemiolo,wta,mslaz,madrian,jobczyk,kluza,bstachur,wpiotr}@agh.edu.pl

Abstract. This paper explores a yet another approach to Explainable Artificial Intelligence. The proposal consists in application of Constraint Programming to discovery of internal structure and parameters of a given black-box system. Apart from specification of a sample of the input and output values, some presupposed knowledge about the possible internal structure and functional components is required. This knowledge can be parameterized with respect to functional specification of internal components, connections among them, and internal parameters. Models of constraints are put forward and example case studies illustrate the proposed ideas.

Keywords: Explainable artificial intelligence · Model discovery · Structure discovery · Model-based reasoning · Causal modeling · Constraint programming

1 Introduction

Discrete Constraint Programming and Discrete Constraint Optimization are inspiring domains for investigation and areas of important practical applications. Over recent decades, wide and in-depth theoretical studies have been carried out, and a number of techniques and tools have been developed [1–3]. A state-of-the-art is summarized in [2]. Unfortunately, the investigated *Constraint Satisfaction Problems* are not only computationally intractable; they are also diversified w.r.t. their *internal structure*, *types of constraints*, and *formalization possibility*. Specialized methods working well for small problems – see e.g. [4] – are not easy to translate to large ones.

The aim of this paper is to sketch out some lines of exploration of *Constraint Programming* (CP) referring to a specific subarea of *Explainable Artificial Intelligence* (XAI), or, more precisely, discovery of internal structure of *Functional Components* (FC) [5] and their parameters within a given system. In fact, the main focus is on an attempt at potential structure discovery and its analysis

M. Stettinger et al. (Eds.): ISMIS 2020, SCI 949, pp. 171–191, 2021.
https://doi.org/10.1007/978-3-030-67148-8_13

for explaining the behaviour of systems, covering a specified collection of input-output data examples. It is assumed that such a system is composed of several FCs belonging to a predefined set of data processing operations. The components are interconnected, and for each of them there are defined several inputs and a single output. Apart from given input-output examples, some, even partial knowledge about the internal structure, is highly required. On the other hand, some component functionalities, parameters, and connections can be discovered.

The application of Constraint Programming seems to be an original contribution in this area. In fact, the content of this paper is located on intersection of Model-Based Reasoning [6], Explainable Artificial Intelligence [7,8], Causal Modeling [9–11], and Constraint Programming [1,2].

This paper reports on work-in-progress; it is a further exploration of the ideas first put forward in [12,13] and [5]. Some related developments and experiments in the area of Constraint Programming application to develop model in the Business Process Modeling (BPMN) we reported in [14–17].

2 Logical Perspective on Explainable Artificial Intelligence

One of the key activities in Artificial Intelligence is based on performing rational inference in order to solve a problem being given some current observations and assuming some background knowledge. Taking into account the classical *logical point of view*, there are three basic logical inference paradigms serving this purpose; these are (i) *deduction*, (ii) *abduction* and (iii) *induction*.

Deduction is a classical inference model widely applied in logic and Artificial Intelligence. It consists on applying defined rules to given facts, so that new knowledge is produced. This paradigm is widely applied in *declarative programming* and *rule-based systems* [18]. It can be considered as application of *self-explainable knowledge* since the definition of rules is typically done by domain experts.

In this paper we follow a yet another approach based on abduction. Abduction consists in finding explanations for current observations (e.g. system behaviour) having some assumed background knowledge. In order to search for the explanation a particular formulation of *Constraint Programming* is put forward.

After [19] let us briefly explain the model of abductive reasoning. Let D denote the *Decision knowledge* (for intuition, some selection of input/decision variables values) and let KB be some available knowledge (the *Knowledge Base*). Let M denote the current *Manifestations* (for intuition, selection of output variables values). The problem of abductive reasoning consists in search for such D, so that:

$$D \cup KB \models M, \tag{1}$$

i.e. D and KB implies (explains) M, and simultaneously

$$D \cup KB \cup M \not\models \bot \tag{2}$$

i.e. the explanations D must be consistent with the observations M in view of available KB. Note that abduction provides only *potential* (but admissible) explanations, and there can be more than one rational explanation in case of realistic systems.

Machine Learning follows in general the pattern of induction. Induction consists in finding universal model covering the set of training examples, so that also new examples can be solved. Note that the logical model of induction can look the same as (1), but now D stays for given input decisions and we search for a general theory KB (e.g. in form of a set of universal, high-level rules) explaining given (a training set) and new (a test set) cases M. Unfortunately, modern approaches to the so-called *Deep Learning* (and classical ones using NN-like or SVM algorithm) produce black-box models and the explanation must be worked out as additional task. An extensive recent survey of such approaches is presented in [20].

Contrary to black-box Machine Learning, abductive approach directly produces explainable, forward-interpretable solutions [5,12,13].

3 An Introductory Example

In this Section, an example illustrating the ideas put forward in this paper are presented in brief. The main goal is twofold: (i) to explain the basic concepts and notions used, and (ii) to show where we are going and what means are used.

Consider a simple *Functional Component* (FC) being a block of several input signals and a single output as presented in Fig. 1. Readers familiar with the *Neural Networks* would recognize perhaps a (part of) simplest model of artificial neuron, but its internal structure is just an example of what can be inside an FC. Let us focus on some essential characteristics important for further discussion.

We analyze an FC with: (i) n input signals $x_1, x_2, \ldots, x_n$, each of them multiplied by an individual weight $w_1, w_2, \ldots, w_n$, and (ii) an internal function defining the way of processing the inputs (in Fig. 1); in our case this is a summation of weighted inputs (typically it may be followed be a threshold function), and (iii) connected to the single output Y. Assume we are given some candidate specifications of the internal structure (and this is our Knowledge Base KB and a set of input-output data records describing the behaviour of the FC; The input values of the x-es and the corresponding output values of Y are our observations M. We are looking for the accurate values of the weights, the definition of the internal function and its parameters, and these are the specification of D. Obviously, the solution must be admissible, i.e. such that all the input-output examples are covered and the produced output values are 100% correct. A discrete, deterministic, finite-domain case is assumed.

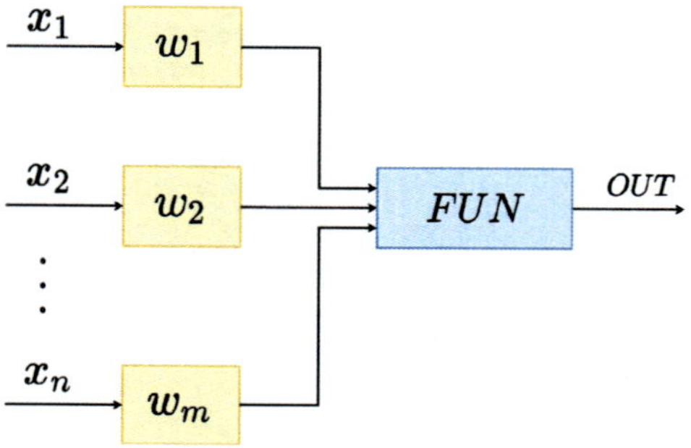

Fig. 1. A Functional Component internal structure

In order to approach the problem, the *Constraint Programming* technology is proposed to search for the model of FC details. We illustrate the example with a simple MiniZinc[1] code excerpt.

```
function var int: wsum(
    array[1..M] of var int: x,
    array[0..M] of var int: w) =
    sum(i in 1..M)(w[i]*x[i]) + w[0];
function var int: wprod(
    array[1..M] of var int: x,
    array[0..M] of var int: w) =
    product(i in 1..M)(w[i]*x[i]) + w[0];

constraint (fun == add) ->
    forall(i in 1..N)(OUT[i] = wsum(INP[i,..],W));
constraint (fun == mult) ->
    forall(i in 1..N)(OUT[i] = wprod(INP[i,..],W));
```

First, we define a set of *candidate functions*; for simplicity, we have just two of them: `wsum` being a weighted sum of the inputs $w_i * x_i$ for $i \in [1..n]$ and a single value `w[0]`; the other one is the `wprod` function being a product of weighted inputs, again with the `w[0]` added to it. To simplify this example only integer values are in use; the variables and parameters definitions are omitted. Since we want to identify *not only the weights* but the *type of the function as well*, we introduce an enumerated type `fun` = {`add`, `mult`}. In the `constraint` definition the trick of the so-called *reification* [5,12] is used: if the Boolean condition `(fun == add)` is true, then the `wsum` function is selected, and if the condition `(fun == mult)` is true, then the `wprod` is accepted. We look for the actual values of the weights and the function.

---

[1] We use the MiniZinc Constraint Programming language: https://www.minizinc.org/ run under Linux Mint.

Now, consider an example case of some *idealized* (but real) solution. Consider a table covering five input vectors specified in as rows of table `INP` and the appropriate outputs given in vector `OUT` as follows:

```
INP = [|3 ,2 ,2 ,3 ,3|
        1 ,2 ,3 ,4 ,5|
        2 ,3 ,3 ,2 ,2|
        3 ,2 ,1 ,2 ,3|
        4 ,1 ,3 ,2 ,3|];
OUT = [12 ,12 ,12 ,14 ,16];
```

The idea is that for the input `[3,2,2,3,3]`, the output must be equal to `12`, etc. After running the solver we obtain *exactly one solution* being the function `wprod` with the weights vector `[-8,3,3,1,-3,4]`. Hence $w_0 = -8, w_1 = 3, \ldots, w_5 = 4$. The weights searched for were limited to the interval $[-10, 10]$.

This example shows the basic idea: *Constraint Programming* can be used for finding *detailed explanation of the component structure, functionality and parameters*. From logical point of view we perform abductive reasoning: we look for possible explanation of the observed behaviour (the type of the functions and the values of the weights). Obviously, as in Constraint Programming, if the input-output data is too poor, a large number of *admissible models* can be generated. Also, if the input-output data is too specific, it may be inconsistent with the assumed class of models, and as a consequence no admissible model would be found.

4 State of the Art in Explainable Artificial Intelligence and Related Work

According to a vast review on the subject [7], **understandability** is a key concept of explainable methods in Artificial Intelligence. Intelligible model is a model, such that its functionality can be comprehended by humans. This particular characteristic is closely tied to **interpretability** and **transparency**. It is also referred to as *white-box models* in Model-Based Reasoning and Systems Science.

According to [7], the first factor is a measure of the degree to which an output can be understood by human. The second trait refers to the inherent internal features of the specific models, i.e. possibility to easily follow the way of data processing leading to generate the output or ease in presentation of results.

This capability can be observed notably among primary Machine Learning techniques like linear regression or tree and rule based models. Nevertheless, with Deep Neural Networks gaining its popularity among researchers and companies, transparency is no longer an asset of the Artificial Intelligence systems. Using *black-box* models created a gap between performance and explainability (Fig. 2) of models that are used in learning [8].

When talking about more sophisticated Machine Learning methods, *post-hoc* explainability is usually applied. It means that some additional techniques are added to the model to make its decisions justified and understandable. According to [7], there are two types of adapted strategies: model-agnostic, which can be

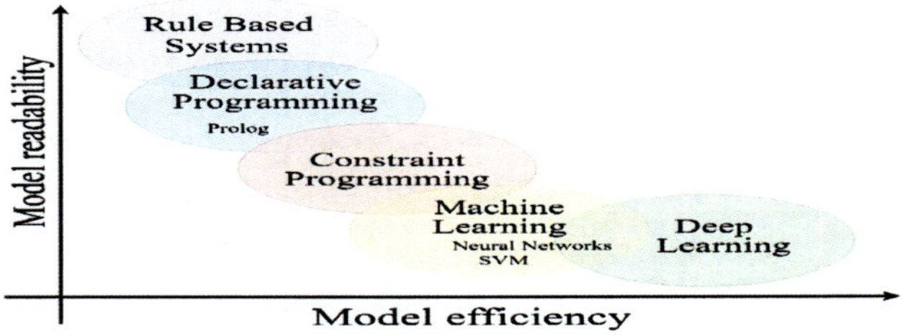

Fig. 2. Trade-off between performance and explainability.

used to any type of Machine Learning model, and model-specific, which must be applied in correspondence with a selected learning tool.

Local Interpretable Model-Agnostic Explanation (LIME) [21] is one of the most widely-used methods. It focuses on building linear models which approximates and simplifies the unintelligible outputs of primary solutions. Additionally, there are SHapley Additive exPlanations (SHAP) [22] which measure certainty calculated for each prediction basing on features relevance in terms of the task. There exists also a branch of methods that base their operation on data visualization. Grad-cam [23] is such a tool that enables users to see which fragment of a photo is most important in terms of a given output, e.g. the head of a dog in the picture recognized as a *dog*.

Apart from photos and time-series, explainability is also used in typical Constraint Programming tasks such as planning. Authors of [24] introduced tools for creating plans that are explicable and predictable for humans. And although there are many tools of this type in explainability, there are still too few in domain of Constraint Programming. Moreover, none of those focuses on increasing the level of explainability through transparency thanks to structure discovery. Finally, note that apart from a plethora of Machine Learning approaches resulting with black-box models, in the area of classical, symbolic (e.g. logic based) AI [25] explainability is a born-in feature; examples include: graph-search algorithms, causal modeling, automated planning, constraint programming, Bayes networks, rule- and expert systems, as well as all founded in logical formalism knowledge bases.

5 Exploring Constraint Programming for Explainable Reasoning About Systems. A Note on Methodological Issues

In classical Machine Learning (ML) one is typically given the input-output training examples (referring to the logical model (1), these are the input decision values D and the observations M), but no knowledge about the internal structure and components (the KB) is provided. Hence, the models created by classical ML – although they cover some or most of the examples – do not explain the rationale behind the observed behavior. So such models are also often referred to as *black-box* ones. For example, recent developments in the so-called *Deep Learning* with large neural nets may exhibit satisfactory results in complex pattern classification, but the hidden knowledge does not undergo any rational, logical analysis and verification. A distinctive example – the Bayesian Networks/Causal Graphs – introduce some insight into the internal structure, but still the induced models are based on probabilistic or qualitative knowledge representation with all the inherent deficiencies.

An important AI area remaining in opposition to the shallow ML models is the so-called Model-Based Reasoning [6]. Areas of application include reasoning and modelling cyber-physical systems, structural reasoning and structure discovery, diagnostic reasoning, causal and qualitative modeling, etc. Techniques such as Causal Modeling [9–11], Consistency-Based Diagnosis [26,27], Abduction [28], or Constructive Abduction [13] might be useful in *white-box* modeling for analysis and explanation of systems behaviour.

Recently, an interest in applying Constraint Programming techniques in subareas of ML can be observed [11,29]. In this paper, we follow the research ideas presented in [5,12,13], also based on use of Constraint Programming tools.

Although the basic problem statement of Constraint Programming (CP) [1] looks very simple, this technology can be used in various ways. Below, with reference to the introductory example presented in Sect. 3, we present a short note on CP methodologies for application in discovery of internal structure and explanatory analysis. The following, simplifying assumptions are accepted:

- discrete, finite valued cases are considered,
- mostly integer and Boolean values are used,
- only exact matching is considered for covering all input cases,
- components – if to be identified – must belong to a predefined selection of them,
- pre-specified knowledge about causal structure is highly desired.

Parameters Values Identification

This is the simplest and most straightforward CP application. Given a set of variables, the constraints are defined with legal constructs of the accepted language (e.g. MiniZinc). The constraints are usually of local nature and reflect the model structure. An example are the weights values in the model of Sect. 3. This approach follows directly the abductive reasoning paradigm defined by Eq. (1).

Parameters Values Identification: Restriction of the Number of Solutions

In case of large number of admissible solutions several auxiliary techniques and constraints can be applied. These include: (i) model decomposition, e.g. by cut-set introduction, (ii) variable domains restriction, (iii) elimination of redundant solutions, (iv) symmetry braking, (v) additional constraint specification and (vi) introducing a quality measure for admissible solution and turning the Constraint Satisfaction Problem in a Constrained Optimization one.

Functional Component Identification

Identifying the function of a FC is not a straightforward task. The proposed solution is to define a finite set of basic FC operations (such as weighted sum of inputs, product, threshold/switching function, sigmoid function, or a polynomial one, etc.), and apply the techniques of *reification*. This requires to define either Boolean variables conditioning the incorporation or exclusion of the specific function or enumerable types of such functions. For example, an FC specification with a 3-order polynomial function can be as follows:

```
function var int: wpol3(
    array[0..3] of var int: a, var int: x) =
    a[3]*pow(x,3) + a[2]*pow(x,2) + a[1]*x + a[0];
constraint (fun == wpol3) -> Y = wpol3(A,X);
```

where `A[0..3]` is the vector of polynomial coefficients searched for, and `X[1..3]` is the current input. The reification means that the function `wpol3` is identified as active only if the predicate `fun == wpol3` takes the value `true`. Note that here we have again the paradigm of abduction defined with Eq. (1), but the set D is extended with extra logical values being a kind of selectors for identification of active functions.

Connection Identification

It is worth noting that in case of structure discovery one can explore constraints defining connection existence or lack of it. Let `X,Y` be two variables (points) to be connected or not. By introducing a Boolean variable `connected_X-Y`, a simple reification constraint can be specified as follows:

```
connected_X-Y -> X = Y;
```

Here `connected_X-Y` is a Boolean variable conditioning the constraint `X = Y` and -> is the *logical implication* symbol; in this way the reification technique is applied. But from logical perspective again the paradigm of abduction defined with Eq. (1) is used, where the set D is extended with extra logical values being a kind of selectors for identification of existing connections.

6 An Example Case Study: Function Identification and Diagnoses Explanation

For further illustration of the proposed ideas let us let us refer to the multiplier-adder system often explored in Model-Based Diagnosis e.g. [26]; it is presented in Fig. 3.

As a first case, let us assume that the system is correct. The investigated problem is to determine the function of each of the five components. An essential excerpt of the MiniZinc model is as follows:

```
constraint fun[1] == add  -> X = A+C;
constraint fun[1] == mult -> X = A*C;
constraint fun[2] == add  -> Y = B+D;
constraint fun[2] == mult -> Y = B*D;
constraint fun[3] == add  -> Z = C+E;
constraint fun[3] == mult -> Z = C*E;
constraint fun[4] == add  -> F = X+Y;
constraint fun[4] == mult -> F = X*Y;
constraint fun[5] == add  -> G = Y+Z;
constraint fun[5] == mult -> G = Y*Z;
```

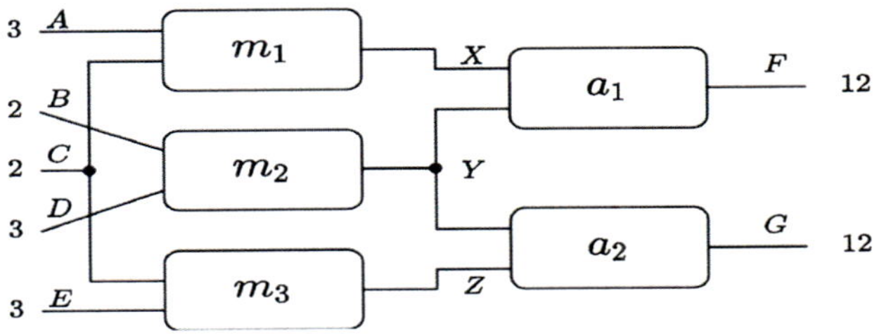

Fig. 3. The multiplier-adder example system. Case of correct operation. [26,30]

For the input values as in Fig. 3 and `F=12,G=12` the functions are correctly identified with the output `fun = [mult,mult,mult,add,add]`. This example shows the potential of Constraint Programming: a single input-output record may be enough to explain the functionalities of five internal Functional Components.

As a second example, consider the case of an internal fault.

Figure 4 presents a system composed of 3 multipliers (located in the first layer) followed by two adders (in the second layer). But now, for the given inputs the output `F=10` is incorrect – it should be `F=12`. The four *logical* potential diagnoses are: $D_1 = \{m1\}$, $D_2 = \{a1\}$, $D_3 = \{a2, m2\}$ and $D_4 = \{m2, m3\}$.

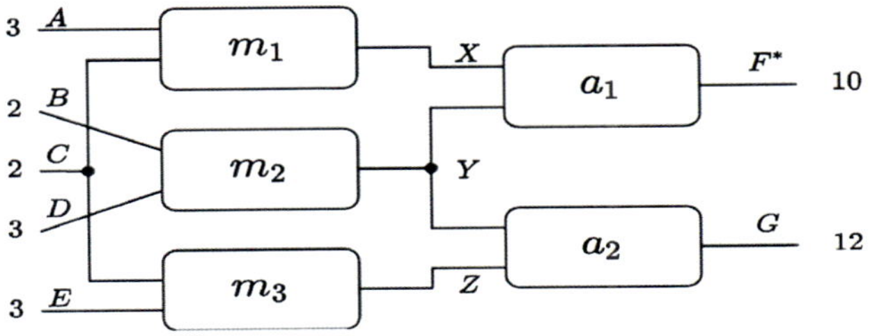

Fig. 4. The multiplier-adder example system. Case of internal fault. [26,30]

Note that only minimal diagnoses are considered. A diagnosis is an explanation stating that the indicated component(s) is (are) faulty. For detailed analysis see e.g. [26]; an even more detailed study introducing *qualitative diagnoses* providing further qualitative information, i.e. if a faulty component lowers (e.g. $a1(-)$; $m1(-)$) or increases the signal value (e.g. $a2(+)$; $m3(+)$ - in the 2-element diagnoses $\{a2(+), m2(-)\}$; $\{m2(-), m3(+)\}$) is presented in [30].

We shall build 4 constraint models for the 4 diagnostic cases for the observations presented in Figure 4. The ultimate goal is to obtain detailed, numerical characteristics of the faulty component behavior i.e. explanation of the fault.

Consider the first case of `m1` being faulty; this is stated as `(not m1)`, where `not` denotes the *logical negation*. We assume that a faulty multiplier produces the incorrect output and its value can be expressed as multiplication of the correct value by a factor `kAC/kX`, where both the numbers are integers. The essential model takes the following form:

```
1 constraint (not m1) ->
2     kX*X = kAC*A*C /\
3     Y = B*D /\  Z = C*E /\ F = X+Y  /\   G = Y+Z;
```

The first constraint (line 2) models the fault of `m1`; the symbol /\ stays for *logical conjunction*. The other constraints correspond directly to the normal operation and connections of the components. The produced output is: `X=4, Y=6, Z=6, kAC=2, kX=3` and its correctness can be easily checked by hand. The values of `kAC=2, kX=3` explain the numeric characteristics of the fault in detail.

As the second case, consider the diagnosis of `a1` being faulty; this is stated as `(not a1)`. We assume that this time it is the adder that produces the incorrect output and its value can be expressed as subtraction from the correct value a factor `kF` (an integer). The model takes the following form:

```
1 constraint (not a1) ->
2     X = A*C     /\   Y = B*D   /\ Z = C*E /\
3     kF+F = X+Y /\   G = Y+Z;
```

The fourth constraint (line 3) models the fault of `a1`. The other constraints correspond directly to correct operation and connections of the components. The produced output is `X=6, Y=6, Z=6, kF=2` and again, it can be easily checked by hand.

The third, a bit more complex case is the one of active diagnosis {`a2,m2`}; this is denoted as `((not m2 /\ not a2))`. The model for this case is as follows:

```
1 constraint ((not m2) /\ (not a2)) ->
2     X = A*C /\  kY*Y = kBD*B*D /\  Z = C*E /\
3     F = X+Y /\  G = Y+Z+kG;
```

Variables kBD/kY model the multiplicative fault of m2 and variable kG models the fault of a2. Note that the double fault exhibits the effect of *compensation* at the output G=12 which is correct. The produced output is X=6, Y=4, Z=6, kBD=2, kY=3, kG=2 and again, it can be easily checked by hand.

The fourth, and perhaps the most complex case, is the one of active diagnosis {m2,m3}; this is denoted as ((not m2 /\ not m3)). Here we have to introduce four variables, namely kBD/kY and kCE/kZ for modeling two multiplicative faults, namely the one of m2 and m3, respectively. The model is as follows:

```
constraint (not m2) /\ (not m3) ->
    X = A*C /\  kY*Y = kBD*B*D /\ kZ*Z = kCE*C*E /\
    F = X+Y /\  G = Y+Z;
```

The produced output is X=6, Y=4, Z=8, kBD=2, kY=3, kCE=4, kZ=3 and again, it explains the numerical details of the fault and can be checked by hand.

In the above example we used the simplest methodology for constraint modeling applied to solving parametric identification problems of finding *detailed numerical explanations* – in fact *numerical models of faulty behavior* – for all the four *minimal diagnoses*.

Note that the presented approach for explainable reasoning with Constraint Programming can be specified as a generic procedure as follows:

- take a minimal diagnosis for detailed examination; it can be composed of k faulty components,
- for any component define one or more (as in the case of multipliers) variables with use of which one can capture the idea of the misbehavior of the component,
- define the domains of the variables (a set necessary in CP with finite domains),
- define the constraints imposed on these variables for the analyzed case, and finally
- define the constraints modeling the possible faulty work of all the components (keep the correct models for components that work correct),
- run the constraint solver and analyze the results.

7 An Extended Example: Practical Structure Discovery with Constraint Programming

In this section the proposed methodology will be step-by-step explained and illustrated with a more practical example. This is still a rather simple and well-known system; for the investigation we have selected the controller (in fact the BCD decoder) of the 7-segment display[2]. The presentation is focused on methodological issues. Despite the simplicity of the selected example, the presentation should assure the satisfaction of the following ultimate goals:

[2] See: https://en.wikipedia.org/wiki/Seven-segment_display.

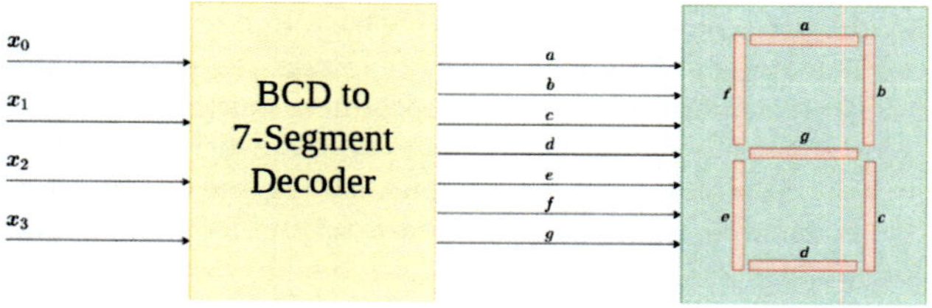

Fig. 5. An example scheme presenting the input and output of the controller and connection to the 7-segment display

- the potential of application of Constraint Programming to structure discovery should be made clear,
- with relatively simple formalization the presentation of the methodology should remain transparent,
- finally, the resulting structure is both readable and verifiable; it explains the behaviour and assures the correct work of the system controller.

The overall idea of the system is presented in Fig. 5. The controller under interest remains a black-box with 4 binary input signals and 7 binary output signals. It is assumed that the input signals encode the decimal numbers `0,1,2,3,4,5,6,7,8,9` which should appear on the display by activating an appropriate combination of the 7 segments `a,b,c,d,e,f,g`.

The inputs of the controller are the binary signals x_3, x_2, x_1, x_0. The digit `D` to be displayed is calculated in and obvious way:

$$D = x_3 * 2^3 + x_2 * 2^2 + x_1 * 2 + x_0$$

and is consistent with simple binary encoding of decimal numbers.

Now, let us present the specification of the given inputs and related to them outputs. The input table `INP` has 4 columns representing the values of x_3, x_2, x_1, x_0, respectively, and 10 rows, each of them representing a single digit of `0,1,2,3,4,5,6,7,8,9`:

Input Data

```
INP = [|0,0,0,0|
        0,0,0,1|
        0,0,1,0|
        0,0,1,1|
        0,1,0,0|
        0,1,0,1|
        0,1,1,0|
        0,1,1,1|
        1,0,0,0|
        1,0,0,1|];
```

The output table `OUT` has 7 columns, each of them corresponding an appropriate display segment `a,b,c,d,e,f,g`, and exactly 10 rows, again each of them representing a single digit of `0,1,2,3,4,5,6,7,8,9`:

Output Data

```
OUT = [|1,1,1,1,1,1,0|
        0,1,1,0,0,0,0|
        1,1,0,1,1,0,1|
        1,1,1,1,0,0,1|
        0,1,1,0,0,1,1|
        1,0,1,1,0,1,1|
        1,0,1,1,1,1,1|
        1,1,1,0,0,0,0|
        1,1,1,1,1,1,1|
        1,1,1,1,0,1,1|];
```

Recall, that the given specification is *an exact one*; the discovered controller structure must assure a 100% compatibility with the desired behaviour. On the other hand, it must remain transparent, and a rational requirement is that it should be as simple as possible.

When searching for the internal structure of the controller we shall use some auxiliary and arbitrary knowledge and decisions. Some background knowledge on Propositional Logic and the design and working principles of digital circuits would be helpful. In fact, we assume the following restrictions on our project:

- we shall look for a structure corresponding to the *Disjunctive Normal Form* (DNF), which is perhaps the most popular in digital circuits design,

- there will be two levels of components, the first one corresponding to AND gates (binary multiplication) and the second one corresponding to OR-gates (binary addition); the negations of input signals are assumed to be available and for simplicity no third level for representing this operation is introduced,
- the number of the AND-gates should be minimal,
- The number of the OR-gates is equal to 7; this is the consequence of the fact that there are 7 independent outputs a,b,c,d,e,f,g,
- once again, the function of the controller must reconstruct the specified input-output behavior in an exact way.

Now, what are the search areas open for the Constraint Programming? These are the following:

- first, we search for connections, from the input signals x_3, x_2, x_1, x_0 and their negations to the inputs of the AND-gates; note that the case of the lack of the connection (if the signal is not used) must also be represented,
- second, we insist that the number of connections should be minimal; this is so to keep the transparency and an action towards uniqueness of the solution,
- third, we want to keep the number of the AND-gates as small as possible, and, finally
- if some AND-gates happen to appear more than once, only one physical realization is necessary.

Let us follow the investigation by explaining step-by step the formalization of principal search components Constraint Programming language being the MiniZinc.

In order to make to code readable, let us start with the following short list of necessary data declarations:

Program Code

```
par int: N; %&% number of input rows
par int: M; %&% number of AND gates in the first layer
array[1..N,1..4] of par 0..1: INP; %&% INPut table
array[1..M,1..4] of var -1..1: W; %&% Weights table
array[1..N,1..7] of par 0..1: OUT; %&% OUTput table
var int: Y; %&% cost function = number of connections
par 1..7: S; %&% segment number; a=1,b=2,..., g=7
```

All the parameters are explained by the appropriate comments. The exact INP and OUT tables have been presented in this section. Both of them have N = 10 rows as there are exactly 10 patterns of required behavior of the controller. M is a parameter searched for each of the 7 output functions: this is the number of the AND-gates and it should be kept minimal. Y is the number of connections for a specific output function.

The key element here is the array W encoding the existence (or not) of the input connections for each of the AND-gate. We applied the following encoding:

- `W[i,j] = 1` codes a connection of (positive occurrence) of x_j to an input of the `i`-th `AND`-gate,
- `W[i,j] = -1` codes a connection of negative occurrence (logical negation) of x_j to an input of the `i`-th `AND`-gate,
- `W[i,j] = 0` codes a lack of connection of x_j as an input of the `i`-th `AND`-gate; in fact, the value of x_j is replaced with a constant `1` (a logical trick for assuring the correct operation of the `AND`-gate).

The corresponding function is a function taking as the input the appropriate x_j value and depending on the value of `W[i,j]` returns the exact or negated value of x_j, or just `1` in case of lack of the connection. The function itself is specified as follows:

Program Code

```
function var int: foo(var -1..1: w, var 0..1: x) =
                    if w==1 then x
                    elseif  w==-1
                    then abs(x-1)
                    else 1 endif;
```

The next step is to define the function encoding a 4-input `AND`-gate parametrized by the existence (or not) of the connections represented with the array `W`. The function is defined as follows:

Program Code

```
function var int: and4(array[1..4] of var  0..1:x,
                       array[1..4] of var -1..1:w) =
                       product(i in 1..4)
                       (foo(w[i],x[i]));
```

This simple function produces the actual output value for the connected x_3, x_2, x_1, x_0 signals or their negations; in case of lack of a specific connection the value `1` is supplemented.

The final function we need is the function producing each of the 7 desired outputs. It is defined as follows:

Program Code

```
function var int: sumM(array[1..M] of var 0..1: m) =
                  if sum(j in 1..M)(m[j]) > 0 then 1
                  else 0 endif;
```

In fact, this is equivalent of the logical `OR`-gate operating on integer values. If at least one of the inputs is `1`, then the output is also equal to `1`; in the other case it is equal to `0`. Since the number of the inputs (coming from an unknown a priori number of `AND`-gates) is unknown, the function is parametrized by `M`.

Now we need to define the number of input connections to be minimized. It is done with the following expression:

Program Code

```
Y = sum(i in 1..M, j in 1..4)(abs(W[i,j]));
solve minimize Y;
```

The interpretation of the code is straightforward: every non-zero element of the array `W` counts as `1`.

And finally, the key condition is that the system must work: for every input the appropriate output must be rendered. This is achieved with a constraint construction as follows:

Program Code

```
N = 10;
constraint forall(i in 1..N) (OUT[i,S] =
                 sumM(j in 1..M)
                 (and4(INP[i,..],W[j,..])));
```

Again, the interpretation is straightforward: for any (the *forall* constraint) of the 10 input rows, the generated output must be equal to the one provided by the specification of the `OUT` table.

Now, let us present the results and some summary of them. As a first example, consider the solution for the `a`-segment function. The minimal number of `AND`-gates is `M = 4`(for lower values of `M` there is no admissible solution). The minimal solution has 6 connections, and is displayed as:

Output Data: segment a

```
W     = [0,  1,  0,  1,
         0,  0,  1,  0,
         1,  0,  0,  0,
         0, -1,  0, -1]
Y = 6
```

Recall that `0` means no connection, while `-1` means negated value of the appropriate `x`. Taking this into account the discovered function is of the form:

$$a = x_2x_0 + x_1 + x_3 + \bar{x}_2\bar{x}_0$$

where the $\bar{x}$ denotes the negation of x.

For the other 6 functions we have the following results:

Function `b`: `M = 3, Y = 5,`

Output Data: segment b

```
W     = [0,  0,  1,  1,
         0, -1,  0,  0,
         0,  0, -1, -1]
Y = 5
```

and so the function is defined as:

$$b = x_1x_0 + \bar{x}_2 + \bar{x}_1\bar{x}_0.$$

Function `c`: `M = 3, Y = 3,`

Output Data: segment c

```
W     = [0,  0,  0,  1,
         0,  1,  0,  0,
         0,  0, -1,  0]
Y = 3
```

and so the function is defined as:

$$c = x_0 + x_2 + \bar{x}_1.$$

Function d: `M = 5`, `Y = 10`,

Output Data: segment d

```
W    = = [0,  1,  -1,  1,
          0, -1,   1,  0,
          1,  0,   0,  0,
          0,  0,   1, -1,
          0, -1,   0, -1]
Y = 10
```

and so the function is defined as:

$$d = x_2\bar{x}_1x_0 + \bar{x}_2x_1 + x_3 + x_1\bar{x}_0 + \bar{x}_2\bar{x}_0.$$

Function e: `M = 2`, `Y = 4`,

Output Data: segment e

```
W    = = [0,  0,  1, -1,
          0, -1,  0, -1]
Y = 4
```

and so the function is defined as:

$$e = x_1\bar{x}_0 + \bar{x}_2\bar{x}_0.$$

Function f: `M = 4`, `Y = 7`,

Output Data: segment f

```
W    = = [1,  0,  0,  0,
          0,  1, -1,  0,
          0,  1,  0, -1,
          0,  0, -1, -1]
Y = 7
```

and so the function is defined as:

$$f = x_3 + x_2\bar{x}_1 + x_2\bar{x}_0 + \bar{x}_1\bar{x}_0.$$

Function g: `M = 4`, `Y = 7`,

Output Data: segment g

```
W     = [0, -1,  1,  0,
          1,  0,  0,  0,
          0,  1, -1,  0,
          0,  1,  0, -1]
Y = 7
```

and so the function is defined as:

$$g = \bar{x}_2 x_1 + x_3 + x_2 \bar{x}_1 + x_2 \bar{x}_0.$$

To summarize, in the following table we present a juxtaposition of the function parameters and some two simple statistics: the NoTS is the Number of Total Solutions (admissible solutions that are not necessary optimal ones) and the TT is the total time of search in case we search for all solutions. In fact, the search time for the optimal solution is very small in case of this example (for example, in case of segment **d** it amounts to 6.10 s with only 8 solutions explored on the path of improved ones) (Table 1).

The presented examples shows that Constraint Programming can be helpful for solving the problem of structure discovery in case of finite-values discrete systems and in presence of auxiliary, partial structural and functional knowledge.

Table 1. Selected numerical results for the 7-segment display controller reconstruction

Segment	M (No. of AND-gates)	Y (No. of connections)	NoTS	TT [s]
a	4	6	4608	3.53
b	3	5	96	0.07
c	3	3	504	0.11
d	5	10	80640	83.56
e	2	4	12	0.00
f	4	7	8064	1.93
g	4	7	3840	2.31

8 Concluding Remarks and Future Work

The discussion supported with the simple but working examples was aimed at the following observation (or proposal): Constraint Programming used in an appropriate way can constitute a valuable tool for exploring internal structure and parameters of systems composed of Functional Components and contribute to explainable reasoning in Artificial Intelligence.

Some directions of further research include: (i) development of a parameterized set of block functions – potential components of larger systems, (ii) exploring imprecise cases with `solve minimize` of fitness evaluation, and (iii) exploring methods for Causal Graphs discovery and modeling of different types of causal relationships.

For future work, we plan to further improve the logic-based encodings, and also work on a hybrid representation for such problems that would enable to use the advantages of both logic-based approaches and stochastic methods. One of the specialized directions planned to be further explored is the development of BPMN diagrams modeling with Constraint Programming.

References

1. Dechter, R.: Constraint Processing. Morgan Kaufmann Publishers, San Francisco (2003)
2. Rossi, F., van Beek, P., Walsh, T. (eds.): Handbook of Constraint Programming. Elsevier (2006)
3. Hentenryck, P.V., Michel, L.: Constraint-Based Local Search. MIT Press, Cambridge (2005)
4. Ligęza, A.: Polskie Towarzystwo Informatyczne. In: Ganzha, M., Maciaszek, L., Paprzycki, M. (eds.) Proceedings of the Federated Conference on Computer Science and Information Systems 2012, pp. 101–107. IEEE Computer Society Press, Warsaw; Los Alamitos (2012)
5. Ligęza, A.: Polskie Towarzystwo Informatyczne. In: Kryszkiewicz, M., Appice, A., Ślęzak, D., Rybinski, H., Skowron, A., R.Z.W. (eds.) International Symposium on Methodologies for Intelligent Systems, pp. 261–268. Springer, Warsaw (2017)
6. Magnani, L., Bertolotti, T.: Springer Handbook of Model-Based Science. Springer, Heidelberg (2017)
7. Arrieta, A.B., Díaz-Rodríguez, N., Del Ser, J., Bennetot, A., Tabik, S., Barbado, A., García, S., Gil-López, S., Molina, D., Benjamins, R., et al.: Information Fusion (2019)
8. Došilović, F.K., Brčić, M., Hlupić, N.: 2018 41st International convention on information and communication technology, electronics and microelectronics (MIPRO), pp. 0210–0215. IEEE (2018)
9. Pearl, J.: Causality. Models, Reasoning and Inference, 2nd edn. Cambridge University Press, New York (2009)
10. Li, J., Le, T.D., Liu, L., Liu, J.: ACM Trans. Intell. Syst. Technol. **7**(2), 14:1 (2015)
11. Yu, K., Li, J., Liu, L.: A review on algorithms for constraint-based causal discovery. arxiv:1611.03977v1 [cs.ai], University of South Australia (2016)
12. Ligęza, A.: International Conference on Diagnostics of Processes and Systems, pp. 94–105. Springer (2017)
13. A. Ligęza, in *Proceedings of the International joint Conference on Knowledge discovery, Knowledge engineering and Knowledge management, IC3K*, vol. 2-KEOD (SCITEPRESS - Science and Technology Publications, Lisbon, Portugal, 2015), *IC3K*, vol. 2-KEOD, pp. 352–357
14. Wiśniewski, P., Kluza, K., Ligęza, A.: Applied Sciences **8**(9), 1428 (2018)
15. Wiśniewski, P., Ligęza, A.: International Conference on Artificial Intelligence and Soft Computing, pp. 788–798. Springer (2018)

16. Wiśniewski, P., Kluza, K., Jobczyk, B, Stachura-Terlecka, K., Ligęza, A.: International Conference on Knowledge Science, Engineering and Management, pp. 55–60. Springer (2019)
17. Kluza, K., Wiśniewski, P., Adrian, W.T., Ligęza, A.: International Conference on Knowledge Science, Engineering and Management, pp. 615–627. Springer (2019)
18. Ligęza, A, Fuster-Parra, P., Aguilar-Martin, J.: LAAS Report No. 96316 (1996)
19. Ligęza, A., Górny, B.: Springer Handbook of Model-Based Science, pp. 435–461. Springer, Cham (2017)
20. Guidotti, R., Monreale, A., Ruggieri, S., Turini, F., Giannotti, F., Pedreschi, D.: ACM Comput. Surv. **51**(5), 1–42 (2019). https://doi.org/10.1145/3236009
21. Ribeiro, M.T., Singh, S., Guestrin, C.: Proceedings of the 22nd ACM SIGKDD International Conference on Knowledge Discovery and Data Mining, pp. 1135–1144. ACM (2016)
22. Chen, H., Lundberg, S., Lee, S.I.: arXiv preprint arXiv:1911.11888 (2019)
23. Selvaraju, R.R., Cogswell, M., Das, A., Vedantam, R., Parikh, D., Batra, D.:Proceedings of the IEEE International Conference on Computer Vision, pp. 618–626 (2017)
24. Zhang, Y., Sreedharan, S., Kulkarni, A., Chakraborti, T., Zhuo, H.H., Kambhampati, S.: 2017 IEEE International Conference on Robotics and Automation (ICRA), pp. 1313–1320. IEEE (2017)
25. Russell, S., Norvig, P.: Artificial Intelligence: A Modern Approach, 3rd edn. Prentice-Hall, New York (2010)
26. Reiter, R.: Artif. Intell. **32**, 57 (1987)
27. Hamscher, W., Console, L., de Kleer, J. (eds.): Readings in Model-Based Diagnosis. Morgan Kaufmann, San Mateo (1992)
28. Poole, D.: Proceedings of IJCAI-89, Sridharan, N.S. (ed.), pp. 1304–1310. Morgan Kaufmann (1989)
29. Bessiere, C., Raedt, L.D., Kotthoff, L., Nijssen, S., O'Sullivan, B., Pedreshi, D. (eds.) Data Mining and Constraint Programming. Foundations of a Cross-Disciplinary Approach. Lecture Notes in Artificial Intelligence, vol. 10101. Springer International Publishing (2016)
30. Ligęza, A., Kościelny, J.M.: Int. J. Appl. Math. Comput. Sci. **18**(4), 465 (2008)

Deep Distributional Temporal Difference Learning for Game Playing

Frej Berglind[1(✉)], Jianhua Chen[1], and Alexandros Sopasakis[2]

[1] Louisiana State University, Baton Rouge, LA, USA
fbergl1@lsu.edu, cschen@lsu.edu
[2] Lund University, Lund, Sweden
alexandros.sopasakis@math.lth.se

Abstract. We compare classic scalar temporal difference learning with three new distributional algorithms for playing the game of 5-in-a-row using deep neural networks: distributional temporal difference learning with constant learning rate, and two distributional temporal difference algorithms with adaptive learning rate. All these algorithms are applicable to any two-player deterministic zero sum game and can probably be successfully generalized to other settings.

All algorithms in our study performed well and developed strong strategies. The algorithms implementing the adaptive methods learned more quickly in the beginning, but in the long run, they were outperformed by the algorithms using constant learning rate which, without any prior knowledge, learned to play the game at a very high level after 200 000 games of self play.

1 Introduction

In recent years, a number of breakthroughs have been made in computer game playing by combining reinforcement learning with deep neural networks, especially since the success of the AlphaGo [13] system in winning the Go game against the world champion in 2016.

One of the most common and successful methods for reinforcement learning in game playing is temporal difference learning (TD) [16]. It is a group of reinforcement learning algorithms applicable to a wide range of problems. However, most real world problems require more data efficient learning and better performance than what is currently possible with reinforcement learning [7]. The classic approach to TD is learning a strategy by approximating the expected reward, but recent research [1] has shown greatly improved results using distributional TD, which approximates the distribution of the reward.

The goal of this project is to study temporal difference learning algorithms combined with deep neural networks and explore the possible advantages of distributional TD under an adaptive learning rate. As a framework for comparing different algorithms, we have chosen the game of 5-in-a-row.

A probabilistic measure of an action's optimality, which served as an adaptive learning rate for distributional temporal difference learning, is proposed in this

M. Stettinger et al. (Eds.): ISMIS 2020, SCI 949, pp. 192–206, 2021.
https://doi.org/10.1007/978-3-030-67148-8_14

paper. In order to create good conditions for the algorithms to learn, we also put some effort in neural network design and hyperparameter tuning and some of these results will also be covered.

This paper is based on Frej Berglind's master's thesis [3].

2 Related Works

Game playing is a classic area of AI research. Games provide a suitable setting to explore different decision making and learning algorithms. Since the dawn of AI research, games such as Checkers, Chess and Go have been common testing ground for AI experiments. In this section, we will provide a short historical and state of the art overview of methods for game playing algorithms.

The earliest example of reinforcement learning in game playing is Arthur Samuel's checkers experiments from 1959, where he constructed a reinforcement learning algorithm that managed to beat him in the game of checkers after 10 h of practice [12]. That is quite an achievement considering the limitations of a 1950's computer.

Gerald Tesauro was the first to develop in 1992 a backgammon algorithm based on TD learning which was able to reach a level of playing equivalent [17, 19] to the best human player at the time. The program combined TD learning through a neural network [21]. In 1997, IBM's chess computer DeepBlue based on highly optimized Minimax search and a handcrafted evaluation function beat the world champion Garry Kasparov [21].

In 2014 Google Deepmind combined deep learning with Q-learning (a kind of TD-learning) and created an algorithm that learned to play several Atari video games from raw images [10]. In 2017, they published the paper "A Distributional Perspective on Reinforcement Learning" [1] introducing an algorithm that combine deep learning and distributional temporal difference learning showing greatly improved performance on Atari games. It is similar to the distributional temporal difference learning algorithm presented in Sect. 4 of this manuscript.

In 2016, Google Deepmind's AlphaGo [13] using a new reinforcement learning algorithm and a deep neural network, beat a professional Go player. Only a year later AlphaGo was able to beat the world's top ranked Go player [21]. Even more importantly a newer version of the algorithm, which was given the name AlphaGoZero [15] greatly outperformed the original AlphaGo while learning the game and without any prior knowledge. Essentially this started the age of game playing machines where humans no longer stood a chance. As a result a generalized version of this algorithm, which was later named AlphaZero [14] reached superhuman performance in Go, Chess and Shogi without any additional tuning for the different games.

Keeping things in perspective, AlphaZero is likely to outperform TD-learning in playing simple games such as for instance 5-in-a-row which we implement in this manuscript. However, the AlphaZero algorithm is specialized on board games and is computationally expensive. TD-learning on the other hand has a wider range of applications and even though it might not beat AlphaZero for playing board games, progress in this field can be very useful as will become clear with the review provided below.

2.1 Temporal Difference Methods Versus State of the Art

In this section we provide more details behind classic methods for learning to predict. This in essence includes methods such as TD, Monte Carlo, dynamic programming and Sarsa. During the presentation we include respective updating mechanisms for each method as well as weaknesses and relative strengths which can better put into perspective their possible areas of application.

We begin by introducing basic notation to be used in order to be make the presentation easier to follow as well as more precise. We use capital letters to denote random variables and small letters to denote the ground truth. We let for instance S denote the state of a system, A the action space and R the reward. As is typically the case π will denote the policy while V is the estimated value function. The parameter α used in many methods below denotes the learning rate.

The simplest TD method is updating as follows,

$$V(S_t) \leftarrow V(S_t) + \alpha[R_{t+1} + \gamma V(S_{t+1}) - V(S_t)]. \tag{1}$$

where the parameter γ denotes the discount rate which essentially keeps track of how far back we look while assessing the value of an update. Note that the learning rate parameter $\alpha > 0$.

As can be seen from this updating mechanism it becomes clear why the method is even called temporal difference: it involves the difference in predictions of the value function V over two states at different times, $V(S_{t+1}) - V(S_t)$ albeit discounted by γ. TD is fully incremental in contrast to typical MC methods which will be described in more detail below.

TD methods are said to bootstrap and sample by looking forward in time in order to learn. Bootstraping means that the update step (1) estimates something based on another estimation. Bootstraping has the benefit of reducing the variance of your estimates. Practically though the effect of bootstraping can be costly since it implies that you would require more samples. On the other hands bootstraping has been shown to be responsible for faster learning and is preferred over, for instance, Monte Carlo methods which we discuss below. Bootstraping however is also sensitive to initial values (for Q or V) used and care must be taken to avoid artifacts such as an initially decreasing error starting to increase in the long term. In short some of the many advantages of TD methods are:

- learn before final outcome (less memory, less peak computation),
- learn even without a final outcome,
- can even learn from incomplete sequences,
- learn new predictions so that we can use them as we go along.

A Monte Carlo (MC) method updates with the rule,

$$V(S_t) \leftarrow V(S_t) + \alpha[G_t - V(S_t)],$$

where G_t denotes the total reward up to time t as discounted by γ. As can be seen above MC also samples like TD in order to learn but it does not bootstrap. MC is designed to look all the way to the end of the game and then provide a return. TD therefore, in this context, can be thought as an one-step method while MC is multi-step method. In that respect MC is analogous to the method TD-λ. TD-λ is another TD algorithm developed by Richard Sutton [17] based on work on TD methods by Arthur Samuel [12]. TD-λ essentially averages the result produced by TD over a number of steps n. There are in fact two implementations of TD-λ. One averages with a forward view while the other averages with a backward view. The parameter $0 \leq \lambda \leq 1$ is called the trace decay parameter and corresponds to bootstraping. A value of $\lambda = 0$ is pure TD while $\lambda = 1$ is pure MC (thus no bootstraping).

Dynamic Programming (DP) was developed by Bellman in the 1950s [2] and is based on mathematical optimization. In DP the update rule is given by,

$$V(S_t) \leftarrow E_\pi[R_{t+1} + \gamma V(S_{t+1})]. \tag{2}$$

DP does not sample but instead looks ahead one-step and then evaluates the expected value of that state. Dynamic programming also bootstraps. We note that in DP the update rule (2) incorporates the expected value over a given policy π. A policy is simply the strategy that an agent (or player) has in mind for winning the game. Formally however a policy is a probability distribution over actions given states. An off-policy method means that you are learning about a policy different than the one you had. An on-policy method means that you are learning more about the policy you are following.

We can turn a TD method into a control method by updating the policy to be greedy with respect to the current estimate. This is what the state-action-reward-state-action or Sarsa algorithm does. The updating mechanism for Sarsa is,

$$Q(S_t, A_t) \leftarrow Q(S_t, A_t) + \alpha[R_{t+1} + \gamma Q(S_{t+1}, A_{t+1}) - Q(S_t, A_t)].$$

Sarsa essentially learns and updates the policy based on actions taken. The Q value represents the possible reward to be received in the next time step. Sarsa is said to be an on-policy algorithm. Sarsa is used in the reinforcement learning area of machine learning.

In contrast Q-learning is an off-policy type TD control method since learning is not guided by the given policy. Q learning is a sample-based point estimation method of the optimal action value function - based on the Bellman equation. The updating mechanism for Q-learning is,

$$Q(S_t, A_t) \leftarrow Q(S_t, A_t) + \alpha[R_{t+1} + \gamma \max_a Q(S_{t+1}, a) - Q(S_t, A_t)].$$

The value of the function Q indicates the quality of a state S_t, action A_t combination. Note that Q-learning uses sampling to experience the world and determine an expected reward for every state.

In general the methods described above perform best for some applications and not so well for some others. For instance MC methods have lower error on past data but higher error on future data when compared to other methods.

On the other hand, based on several results shown in [17] the slowest TD method is still faster than the fastest MC method. Similarly for dynamic programming to be useful you must essentially have all the information available. If you know nothing about the state of the world around you then dynamic programming cannot help.

Errors and their propagation is of critical importance when considering how to apply these methods. The age old question is: should you perform multiple steps of a 1-step method or rather 1 step of a multi-step method? This however, as Sutton commented "is a trap" [17]. Repeating a 1-step method multiple times is exponentially complex and computationally intractable. Furthermore errors pile up. In real life we have imperfect information.

Similarly important is the question of convergence with any of the methods we use. Both TD and MC methods converge under suitable assumptions.

TD learning is widely used in RL to predict a number of things such as future reward, value functions, etc. The classic approach to TD is learning a strategy by approximating the expected reward. Recent research [1] has shown greatly improved results using distributional TD, which approximates the distribution of the reward.

TD learning furthermore is in the core of many other methods such as Q-learning [20], TD-λ [11,12], Deep Q networks [9,10], TD-gammon [19], etc. One of the significant advantages of TD learning is that it is an optional method which can be added to improve performance. As a result it has become ubiquitous [18].

In general most AI algorithms are not scalable. This is mainly attributed to the data availability bottleneck inadvertently created by humans. Most AI algorithms are weakly scalable [17]. As a result supervised learning methods and model-free RL algorithms are weakly scalable and suffer from the availability of data bottleneck mentioned above. Real world problems however require data efficient learning and better performance than what is currently possible with reinforcement learning [7]. TD learning is, in contrast to many other AI methods, fully scalable [17].

3 Background

3.1 The Game of 5-in-a-row

5-in-a-row is a two-player strategy game traditionally played on squared paper. The two players take turns placing markers in an empty cell on the paper. One player uses "X" and the other player uses "O". You win by getting 5 in a row horizontally, vertically or diagonally. If the grid is filled up without anyone having 5-in-a-row the game ends in a tie. An example of a winning position is shown in Fig. 1.

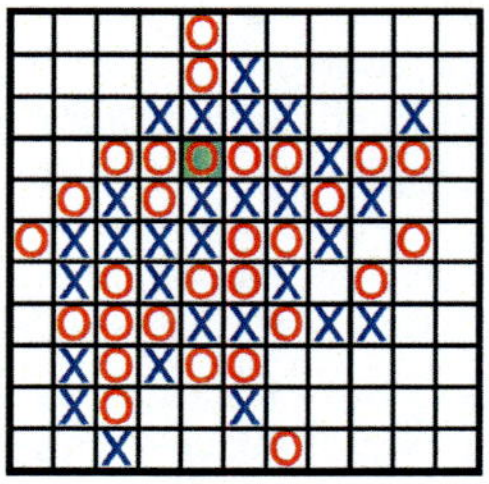

Fig. 1. A game of 5-in-a-row won by "o".

Gomuko is another name for this game. Gomuko is usually played on a 15 by 15 and was proven to be a first player win in 1993 [8]. We have chosen to restrict the game to a board size of 11 by 11. It is large enough to make the game complex while keeping the state and action spaces reasonably small. Since an 11 by 11 board is more restricted than the 15 by 15 Gomuko board, the game should, if played perfectly, either end in with the first player winning or a tie.

3.2 Alternating Markov Games

To define the algorithms in a more general setting, we use an abstract framework for games similar to 5-in-a-row which includes games like Chess, Othello, Checkers and Go. These are *deterministic alternating Markov games* with a final reward of 1 for winning 0 for a tie and -1 for losing and no other reward during the game. An alternating Markov game is similar to a Markov decision process, but differs by having two adversarial agents.

A game is defined by a state space S, an action space $A(s)$ for each state $s \in S$, a transition function $f(s, a)$ defining the successor state when selecting a in s, an initial state s_0 and a function $r(s)$ which determines if a state is final and in that case returns the reward. A final state either has reward 1 (win), 0 (tie), or -1 (loss). The reward for a player is -1 times the reward of the opponent. This symmetric view on the reward is used in this implementation of the algorithms. Another way to think of this is as if one player is trying to maximize the reward and its opponent trying to minimize the same reward.

The game is played by letting the players take turns selecting actions until they reach a final state. A game can be seen as a sequence of states $[s_i]$ connected by actions $[a_i]$ where even numbered actions are played by the first player and the odd numbered actions are played by the opponent. This is illustrated in Fig. 2 and can be the described by the recurrence relation

$$s_{i+1} = f(s_i, a_i), a_i \in A(s_i), \tag{3}$$

saying that the next element in the sequence of states is the successor of the current state depending on which action the agent selects.

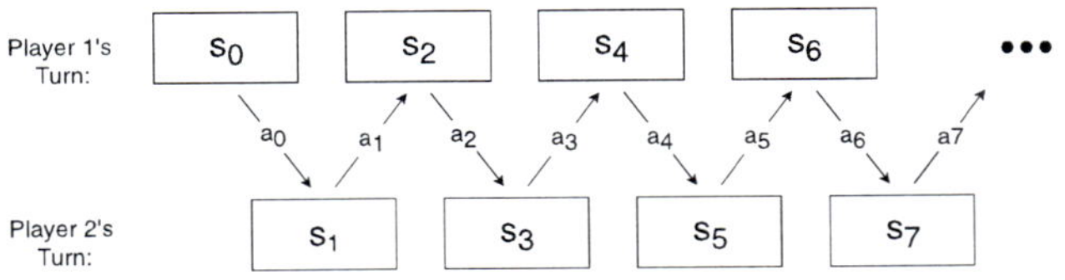

Fig. 2. The game is a sequence of states connected by actions. The states are produced recursively according to Eq. 3. The actions of the first player are the even numbered downward arrows and the moves done by the second player are the odd numbered upwards arrows.

4 Method

In this section, we describe the algorithms starting from basic TD-learning and define the optimality measure used in the new algorithms. Furthermore, we outline how the algorithms are implemented, the training procedure and the evaluation of the results. In Sect. 4.1 to 4.5, for each algorithm, we describe the decision making rule used for optimal play and the updating rule used for training. The algorithms are defined within the framework of the Deterministic Alternating Markov Games described in Sect. 3.2. In order to discover new better strategies, the agents use a suboptimal policy during training, which is described in Sect. 4.6. For a more detailed description of the methods, the reader is referred to the complete thesis [3].

4.1 Temporal Difference Learning

A classic reinforcement learning approach is directly approximating the value function, $V^*(s) \approx V(s)$. This is what is usually referred to as temporal difference learning. In deep temporal difference learning, $V^*(s)$ is calculated by a neural network. The result of this algorithm will serve as a reference point for the other more experimental methods. To distinguish it from the distributional algorithms, we sometimes refer to it as scalar TD. The agent using this algorithm will be called TDBot.

4.1.1 Decision Making

The agent simply selects the move with the highest expected reward,

$$a(s) = \operatorname*{argmax}_{a_i \in A(s)} V^*(f(s, a_i)). \tag{4}$$

4.1.2 Training

After the agent has played a game it uses the result to assign new values to V^*. The update starts from the end of the game by assigning a new value[1] to the final state,

$$V^*(s_{final}) \leftarrow r(s_{final}) \tag{5}$$

and recursively updates V^* based on the Bellman Equation, gradually stepping back through the game,

$$V^*(s) \leftarrow (1 - \alpha)V^*(s) + \alpha\gamma \max_{a_i \in A(s)} \left[V^*(f(s, a_i))\right]. \tag{6}$$

[1] The arrow ($\leftarrow$) is a pseudo code notation for assigning a new value to the function. In our implementation, the new value is used immediately to create new values for preceding states and the input/output pair is used as training data for the neural network at the end of the training iteration.

There are two training parameters: $\alpha \in [0, 1]$ is the learning rate and determines how the agent values old and new knowledge. With $\alpha = 1$ the agent completely discards old knowledge and with $\alpha = 0$ the agent does not change at all. $\gamma \in [0, 1]$ is the discount factor. It determines how the agent prioritize between quick and delayed rewards. $\gamma = 0$ would mean the agent only cares about immediate reward and $\gamma = 1$ makes the agent value all rewards equally.

A state gets updated using its best successor state. When applied to a 2-player game, the value of the successor state will be calculated from the opponent's perspective. In this case, one would use

$$V^*(f(s, a_i)) = -V^*_{\text{opponent}}(f(s, a_i)) \tag{7}$$

to change it to the correct perspective.

The new input-output (state-value) pairs assigned in Eq. 5 and 6 were used as training data for the neural network.

4.2 Distributional Temporal Difference Learning

Distributional Temporal Difference Learning (DTD) is very similar to scalar TD-learning. The only difference is that it learns the distribution of the reward instead of the expectation value. We used a type of DTD where $\gamma = 1$ and the distribution consists of probabilities for ending the game with a win, tie or loss. This discrete distribution is approximated using a neural network,

$$g(s) \approx (p(win|s), p(tie|s), p(lose|s)) \tag{8}$$

where each component of the output vector is the probability of an outcome for the game in state s. The approximated distribution D^* is:

$$\begin{cases} D^*(win|s) &= g(s)_1 \\ D^*(tie|s) &= g(s)_2 \\ D^*(lose|s) &= g(s)_3 \end{cases} \tag{9}$$

In DTD g takes role of V^* as the function which is learned through the training process. The state value can be approximated using g:

$$V(s) = p(win|s) - p(lose|s) \approx g(s) \cdot (1, 0, -1) \tag{10}$$

where "$\cdot$" denotes a dot product.

By creating soft labels with the probability of a win, tie and loss, this algorithm turns the game into a classification problem, unlike the regression problem of learning the continuous values of $V^*(s)$ described in the previous section.

Decision Making

The agent selects the move with the highest expected reward:

$$a(s) = \underset{a_i \in A(s)}{\text{argmax}} \left[g(f(s, a)) \cdot (1, 0, -1) \right] \tag{11}$$

Training
Similarly to TD-learning, DTD uses the result of the game to assign new values to g. The update starts from the end of the game by assigning a new value to the final state,

$$g(s_{final}) \leftarrow \begin{cases} (1,0,0) & \text{for } r(s_{final}) = 1 \\ (0,1,0) & \text{for } r(s_{final}) = 0 \\ (0,0,1) & \text{for } r(s_{final}) = -1 \end{cases} \tag{12}$$

and recursively updates the values according to

$$\begin{cases} a = \text{argmax}_{a_i \in A(s)} \left[g(f(s,a_i)) \cdot (1,0,-1) \right] \\ g(s) \leftarrow (1-\alpha) g(s) + \alpha\, g((f(s,a)) \end{cases} \tag{13}$$

where $\alpha \in [0,1]$ is the learning rate.

To get the correct perspective in a 2-player game you need to reverse $g(f(s,a_i))$ since the successor state will be evaluated from the opponent's perspective and p(player 1 winning) = p(player 2 losing).

4.3 Optimality

An action is optimal if it leads to an optimal outcome from the current state. That is, it is optimal if it leads to a win, it leads to a tie where it is impossible to win, or all actions in the current state lead to a loss. We can calculate the probability of an action being optimal using g. Let $a_i \in A(s)$ and $g(f(s,a_i)) = (w_i, t_i, l_i)$, then

$$p(a_i \text{ optimal}) = p(\text{win}|a_i) + p(\text{tie}|a_i) p(\text{can't win}|a_j, j \neq i) + p(\text{can only lose in } s). \tag{14}$$

$$\iff p(a_i \text{ optimal}) = w_i + t_i \prod_{j \neq i} (1 - w_j) + \prod_j l_j \tag{15}$$

We refer to $p(a_i$ optimal) as the *optimality* of a_i. The optimality shows how strongly connected the values for two consecutive states are. If we choose an optimal action, then the current state and the successor state should have the same ranking. If we make a bad move —choose an action with low optimality— then the current state isn't necessarily bad, but the next is.

4.4 Adaptive Distributional Temporal Difference Learning

The optimality can be used as an adaptive learning rate for DTD. By using $\alpha = p(a(s)$ optimal), we get a new algorithm with adaptive learning rate. We call the agent using this algorithm ADTDBot.

4.5 Original Algorithm

With optimality as an adaptive learning rate, it is no longer necessary to learn from the best successor state. It is possible to simply use the actions and successor states from the game for training. We will refer to this algorithm as BerglindBot.

The decision making is performed in the same way as in DTD. The agent is trained using the following update rule,

$$g(s) \leftarrow (1 - p(a \text{ optimal}))g(s) + p(a \text{ optimal})g(f(s, a)) \tag{16}$$

where a is the action selected in state s during the game.

4.6 Exploration

If the agent always selects the best action it might not discover new and potentially better policies. Furthermore, a neural network needs diverse data to learn well and an agent using its best policy is likely to play quite repetitively. We devised a simple method for adding a good amount of randomization without completely ignoring prior knowledge. It is somewhat similar to the common epsilon greedy method. However, while epsilon greedy completely randomizes a small part of the actions, this method adds a slight randomization to each action.

Let $\hat{V}(s)$ be the approximated expected reward, either calculated directly using $V^*(s)$ or as $g(s) \cdot (1, 0, -1)$. Let

$$\hat{M} = \max_{a_i \in A(s)} \hat{V}(f(s, a_i)). \tag{17}$$

Randomly select an action in the set

$$A_{selected} = \{a_i \in A(s) | \hat{M} - 2\epsilon \leq \hat{V}(f(s, a_i)\} \tag{18}$$

where $\epsilon \in [0, 1]$ is the exploration parameter.

If $\epsilon = 0$, the agent will, according to its current knowledge, select the best possible move. If $\epsilon = 1$, the agent will play completely randomly. If $\epsilon \in (0, 1)$, the agent should play somewhat randomly, but avoid making any critical mistakes. As the training goes on the agent will get better at distinguishing good and bad moves and it should play less and less randomly.

If we assume the approximation error for $\hat{V}$ is less than ϵ,

$$\left|\hat{V}(s) - V(s)\right| \leq \epsilon \qquad \forall s \in S \tag{19}$$

then the best action will be in $A_{selected}$.

4.7 The Opponent

We have previously designed a 5-in-a-row agent which uses heuristics for decision making and this algorithm served as a static reference point during the training. It creates a ranking for each cell of the board based on how good opportunities the players have and selects the cells with the highest ranking. It is fast, systematic and plays the game at the level of an intermediate player as will be shown in the subsequent results. We refer to this agent as QuickBot.

4.8 Training Process

The training is performed as a sequence of training iterations. Each training iteration consists of 3 steps:

1. **Training Games.** Each iteration starts with N (usually set to 50) practice games played using the exploration policy described in Sect. 4.6. After each game the end result is used to assign a new output value to the states in the game. This is done by iterating back from the final state using the training formulas for the different algorithms. The new state/output pairs are saved for training the neural network.
2. **Network Update.** The network is trained for five epochs using the data generated in the previous step.
3. **Evaluation Games.** The agent, using its optimal policy ($\epsilon = 0$), plays 10 evaluation games against QuickBot starting with 1 random move. This is only done to track the progress and the data is not used for training.

Most agents have been trained for 1000 iterations of 50 games. Throughout the training the score and length of each game is saved and this can be used to analyse the learning process.

4.9 Implementation

The players analysed the board using convolutional neural networks. The input of the network was divided to 2 channels, just like two colors in an image. The first channel represented the board positions of the player who placed the most recent move and the second channel represented the board positions for the opponent. Both channels consisted of binary 11 by 11 arrays where a cell is 1 if the player has a marker in this position and 0 otherwise. A similar input representation was used in AlphaGo [13] and its successors.

Five different convolutional network architectures were used:

Small: Convolutional network with 7 layers and 170 000 parameter.
Deep: Convolutional network with 9 layers and 400 000 parameters.
Wide: Convolutional network with 7 layers and 500 000 parameters.
Res: Residual network [5] similar to that used in AlphaZero [14]. It has 14 layers and 380 000 parameters. All experiments presented in this paper except those in Fig. 3a used this network architecture.

Res $\gamma = 0$: The same residual network with batchnorm initialized to zero in accordance to [4].

Res 2: The same as the first resnet, except certain layers are in different order to follow the ResNet2 architecture as proposed in [6].

To make decisions in the game, each possible action in the current state was evaluated by analysing the successor state using the neural network and this data was saved for generating new training data. When a game was finished, new training data was generated by iterating backwards through the game using the update formulas presented in Sect. 4.1 to 4.5. To get eight times more training data, the input is rotated and reflected according to the symmetries of the game. This is the only knowledge about the game, apart from the rules, used by the agents.

TDBot was trained using the Mean Squared Error (MSE) loss function and the discount factor $\gamma = 0.99$. The other algorithms used categorical cross entropy loss. All networks were trained using an Adam optimizer with the common default parameters $\alpha = 0.001, \beta_1 = 0.9, \beta_2 = 0.999$.

5 Results

Throughout the project, we performed many different experiments. Here, we will summarize the most significant results.

Figure 3a shows a comparison between the different network architectures trained with the ADTD algorithm. Since the residual architecture from AlphaZero [14] achieved the best results, it was used for the other experiments. In Fig. 3b, we compare the performance of the different algorithms when training against QuickBot. ADTD and Berglind learned faster in the beginning, but all agents reached a good result in the end. In Fig. 3c, ADTDBot and TDBot were trained through self play using a similar set up as in Fig. 3b. They achieve similar performance against QuickBot, but the symmetry and free exploration of self play stabilize the training and makes it generalize far better against new opponents.

In Fig. 3d, the agents play twice as many games per iteration and train for a total of 200 000 games. This allows us to see the convergence behavior of the algorithms and compare their final performance. By comparing Fig. 3c and 3d, we can see that training with larger data sets makes the agents develop better strategies; all agents except BerglindBot reached an average score above 0.9. ADTD learned faster in the beginning, but gets outperformed by DTD and TD in the end. DTD had a slightly higher score than TD. To further compare the performance of the different agents, two tournaments were played between them. The result is shown in Table 1. It confirms that, in the end, all agents learned to outperform QuickBot, but DTD and TD performed the best.

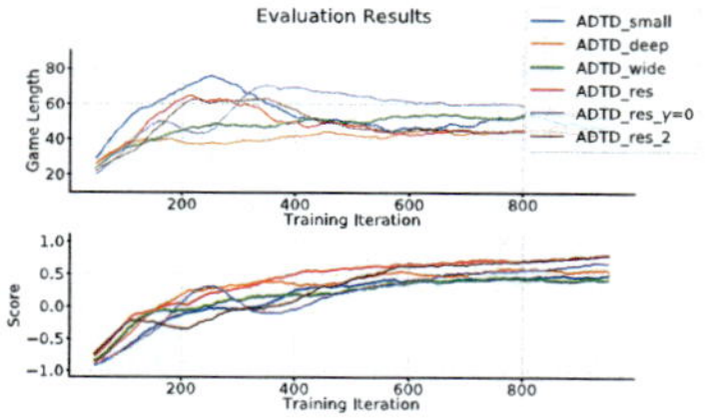

(a) ADTDBot trained against QuickBot for 1000 iterations of 50 games using different network architectures and exploration $\epsilon = 0.01$.

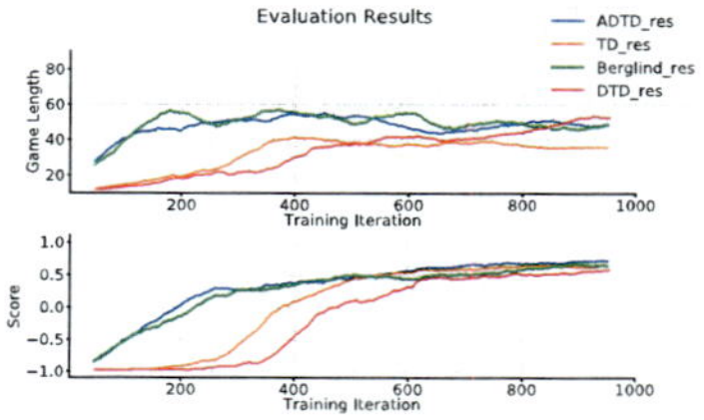

(b) All algorithms trained against QuickBot for 1000 iterations of 50 games using residual networks and exploration $\epsilon = 0.01$. TDBot and DTDBot used the learning rate $\alpha = 0.3$.

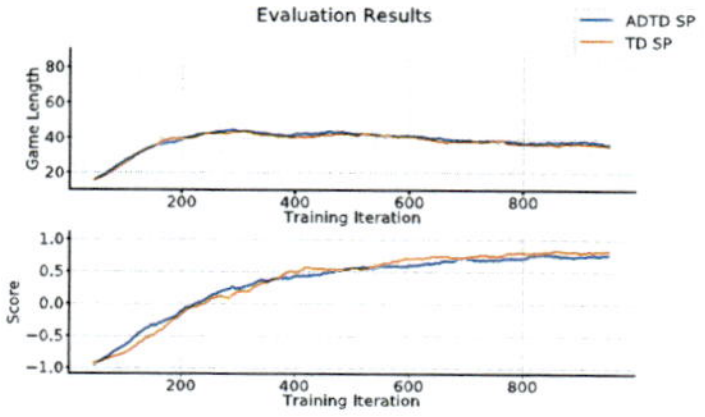

(c) ADTDBot and BerglindBot trained through self play for 1000 iterations of 50 games using ResNet and $\epsilon = 0.05$. TDBot used the higher learning rate $\alpha = 0.8$ to accommodate to the new learning environment (self play).

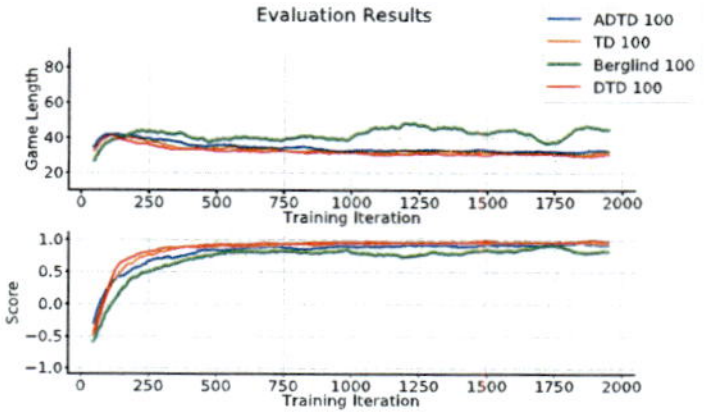

(d) All algorithms trained through self play for 2000 iterations of 100 games using residual networks and exploration $\epsilon = 0.05$. TDBot and DTDBot had the learning rate $\alpha = 0.8$.

Fig. 3. Results of the evaluation games played at the end of each training iteration using the current best policy ($\epsilon = 0$). These games were only played to track the progress and the data was not used for training. The graphs have been smoothed using a running average over 100 iterations.

Table 1. Results from tournaments comparing the performance of different algorithms.

(a) Total score from a tournament with 2000 games played between each pair of players. Each game started with four random moves. These games were played when the training was finished and the data was not used to train the agents	
TDBot Final	2416
DTDBot Final	2173
ADTDBot Final	972
BerglindBot Final	705
QuickBot	−6266

(b) Total score from a tournament with 400 games between each pair of players. Each game started with one random move. These games were played when the training was finished and the data was not used to train the agents	
DTDBot Final	655
TDBot Final	533
ADTDBot Final	272
BerglindBot Final	−29
QuickBot	−1431

6 Conclusions

In this paper, we review temporal difference learning in the context of two player board games. We specifically considered four different distributional temporal difference learning methods. We then trained these algorithms to learn via deep neural networks in order to evolve a winning agent for the game of five-in-a-row.

One important conclusion is that the residual network architecture was most effective in learning, as seen in results in Fig. 3b. Therefore, we adopted this architecture for all our subsequent tests and comparisons for our four TD-algorithms. All four methods performed well and developed strong strategies, but there were significant differences in their performance. Below, are some of the main takeaways.

- **Distributional TD.** The experiments indicate that DTD can achieve strong results, but it is more sensitive to tuning and the neural network capacity. The classic scalar TD works well and is easier to implement and tune than DTD. If one wants quick working result, then scalar TD is a good choice. For the best result and to gain more information, it can be worth experimenting with DTD.
- **Adaptive Learning Rate.** Using the optimality of the selected action as an adaptive learning rate in temporal difference learning leads to more consistent results without parameter tuning. It learns more quickly in the beginning, but does not converge as nicely as DTD with well tuned constant learning rate and has lower performance in the end. More research is needed to evaluate the usefulness of the optimality measure and an adaptive learning rate.
- **Learning From the Best Successor State.** According to our results, it is very beneficial to learn from the best possible successor state, instead of simply using the successor state visited in the game. It allows the algorithm to explore options even though they were not actually played and helps the it learn more quickly and reach a better end result. This benefit is clear when you see how ADTDBot outperformings BerglindBot.
- **Self Play.** Self play helps stabilizing the learning process and lets the agents freely explore the game from the perspective of both players. It leads to an overall stronger strategy and better generalization against new opponents.

References

1. Bellemare, M.G., Dabney, W., Munos, R.: A distributional perspective on reinforcement learning. CoRR abs/1707.06887 (2017). http://arxiv.org/abs/1707.06887
2. Bellman, R.: The theory of dynamic programming. Bull. Am. Math. Soc. **60**(6), 503–516 (1954)
3. Berglind, F.: Deep distributional temporal difference learning for game playing. Master's thesis, Lund University (2020)
4. Goyal, P., Dollár, P., Girshick, R.B., Noordhuis, P., Wesolowski, L., Kyrola, A., Tulloch, A., Jia, Y., He, K.: Accurate, large minibatch SGD: training imagenet in 1 h. CoRR abs/1706.02677 (2017). http://arxiv.org/abs/1706.02677

5. He, K., Zhang, X., Ren, S., Sun, J.: Deep residual learning for image recognition. arXiv preprint arXiv:1512.03385 (2015)
6. He, K., Zhang, X., Ren, S., Sun, J.: Identity mappings in deep residual networks. CoRR abs/1603.05027 (2016). http://arxiv.org/abs/1603.05027
7. Irpan, A.: Deep reinforcement learning doesn't work yet (2018). https://www.alexirpan.com/2018/02/14/rl-hard.html
8. Ailis, L.V., van den Herik, H.J., Huntjens, M.H.: GoMoku solved by new search techniques. AAAI Technical Report FS-93-02 (1993). https://www.aaai.org/Papers/Symposia/Fall/1993/FS-93-02/FS93-02-001.pdf
9. Mnih, V., Kavukcuoglu, K., Silver, D., Graves, A., Antonoglou, I., Wierstra, D., Riedmiller, M.: Playing atari with deep reinforcement learning, NIPS Deep Learning Workshop (2013). arxiv:1312.5602
10. Mnih, V., Kavukcuoglu, K., Silver, D., Rusu, A.A., Veness, J., Bellemare, M.G., Graves, A., Riedmiller, M., Fidjeland, A.K., Ostrovski, G., Petersen, S., Beattie, C., Sadik, A., Antonoglou, I., King, H., Kumaran, D., Wierstra, D., Legg, S., Hassabis, D.: Human-level control through deep reinforcement learning. Nature **518**(7540), 529–533 (2015). https://doi.org/10.1038/nature14236
11. Russell, S., Norvig, P.: Artificial Intelligence: A Modern Approach, 3rd edn. Prentice Hall Press, Upper Saddle River (2009)
12. Samuel, A.L.: Some studies in machine learning using the game of checkers. IBM J. Res. Dev. **3**(3), 210–229 (1959). https://doi.org/10.1147/rd.33.0210
13. Silver, D., Huang, A., Maddison, C.J., Guez, A., Sifre, L., van den Driessche, G., Schrittwieser, J., Antonoglou, I., Panneershelvam, V., Lanctot, M., Dieleman, S., Grewe, D., Nham, J., Kalchbrenner, N., Sutskever, I., Lillicrap, T.P., Leach, M., Kavukcuoglu, K., Graepel, T., Hassabis, D.: Mastering the game of go with deep neural networks and tree search. Nature **529**, 484–489 (2016)
14. Silver, D., Hubert, T., Schrittwieser, J., Antonoglou, I., Lai, M., Guez, A., Lanctot, M., Sifre, L., Kumaran, D., Graepel, T., Lillicrap, T., Simonyan, K., Hassabis, D.: Mastering chess and shogi by self-play with a general reinforcement learning algorithm (2017)
15. Silver, D., Schrittwieser, J., Simonyan, K., Antonoglou, I., Huang, A., Guez, A., Hubert, T., Baker, L.R., Lai, M., Bolton, A., Chen, Y., Lillicrap, T.P., Hui, F., Sifre, L., van den Driessche, G., Graepel, T., Hassabis, D.: Mastering the game of go without human knowledge. Nature **550**, 354–359 (2017)
16. Silver, D., Sutton, R.S., Müller, M.: Temporal-difference search in computer go. Mach. Learn. **87**(2), 183–219 (2012). https://doi.org/10.1007/s10994-012-5280-0
17. Sutton, R.S.: Learning to predict by the methods of temporal differences. Mach. Learn. **3**, 9–44 (1988)
18. Sutton, R.S., Barto, A.G.: Reinforcement Learning: An Introduction, 2nd edn. The MIT Press, Cambridge (2018). http://incompleteideas.net/book/the-book-2nd.html
19. Tesauro, G.: Temporal difference learning and TD-Gammon. Commun. ACM **38**, 58–68 (1995)
20. Watkins, C.: Learning from delayed rewards. Ph.D. thesis, Cambridge University (1989)
21. Yannakakis, G.N., Togelius, J.: Artificial Intelligence and Games. Springer, Cham (2018). http://gameaibook.org

Author Index

M. Stettinger et al. (Eds.): ISMIS 2020, SCI 949, pp. 207–208, 2021.
https://doi.org/10.1007/978-3-030-67148-8